T0261828

Orchids of
Tropical America

Orchids of Tropical America

AN INTRODUCTION AND GUIDE

Joe E. Meisel

Ronald S. Kaufmann and Franco Pupulin

Foreword by Phillip J. Cribb

Comstock Publishing Associates
a division of
CORNELL UNIVERSITY PRESS
ITHACA AND LONDON

First published 2014 by Cornell University Press
First printing, Cornell Paperbacks, 2014

Printed in China

Library of Congress Cataloging-in-Publication Data

Meisel, Joe E., author.
  Orchids of tropical America: an introduction and guide / Joe E. Meisel,
Ronald S. Kaufmann and Franco Pupulin ; foreword by Dr. Phillip J. Cribb.
    pages cm
  Includes bibliographical references and index.
  ISBN 978-0-8014-5335-9 (cloth : alk. paper)
  ISBN 978-0-8014-7768-3 (pbk. : alk. paper)
  1. Orchids—Latin America. 2. Orchids—Tropics. I. Kaufmann,
Ronald S., author. II. Pupulin, Franco, author. III. Title.
  QK495.O64M43   2014
  584'.4098—dc23   2014002084

Cornell University Press strives to use environmentally responsible
suppliers and materials to the fullest extent possible in the publishing
of its books. Such materials include vegetable-based, low-VOC inks
and acid-free papers that are recycled, totally chlorine-free, or partly
composed of nonwood fibers. For further information, visit our website
at www.cornellpress.cornell.edu.

Cloth printing   10 9 8 7 6 5 4 3 2 1
Paperback printing   10 9 8 7 6 5 4 3 2 1

Photos by AWZ Orchids; Fred Clarke, Sunset Valley Orchids;
Ecuagenera; Ron Kaufmann; Valérie Léonard; Joe Meisel; Ron Parsons,
www.flowershots.net; Andy Phillips, www.andysorchids.com; and Juan
Carlos Uribe

All line drawings by Kathy Creger

To resolute people throughout the tropics,
wealthy in land but poor in silver,
who sacrifice and toil to
safeguard orchid-rich forests.

These kinds of Orchis have not bin much written of by the Ancients . . .
but they are chiefly regarded for the pleasant and beautiful floures,
wherewith Nature hath seemed to play and disport her selfe.
—JOHN GERARDE, *The Herball or Generall Historie of Plantes* (1633)

Those who dwell, as scientists or laymen, among the beauties and
mysteries of the earth are never alone or weary of life.
—RACHEL CARSON, *The Sense of Wonder* (1965)

Earth laughs in flowers.
—RALPH WALDO EMERSON, *Hametraya* (1846)

Contents

Foreword

In the spring of 1971, I was fortunate enough to visit the ruins of Machu Picchu in Peru in the company of César Vargas, best known as a prolific contributor to the Flora of Peru project and an expert on the native orchids, which he collected for Charles Schweinfurth, the author of the Flora's orchid account. My outstanding memory of the trip was walking with him up the zigzag road to the ruins and being astounded by the quantity and diversity of orchids growing along the roadside. These impressions were colored by my total lack of knowledge of tropical American orchids, at the time my focus being on the family Solanaceae. Fortunately, César, a walking encyclopedia, identified orchids and pointed out rarities as we strolled along. Most visitors to the tropical Americas do not have an experienced orchid specialist to call upon, and the result can be confusion and discouragement.

We now know that the tropical American orchid flora is the most diverse in the world, with nearly 4000 species having been identified in Ecuador alone. Colombia is likely to be even richer but is less well explored, while even small Costa Rica has well over 2000 species. Contrast this with the orchid floras of Europe at less than 200 species, and of North America, excluding Florida, which numbers less than 300 species. Not only are there copious numbers of species in the American tropics, but the diversity of ecology, size, and form is amazing. They can grow in a remarkable variety of habitats. Orchids are found from the coast to alpine meadows. Some are found in grassland or on the floor of the forest, yet more grow on rocks or trees. Remarkably, a few even grow in streams and rivers or on the edges of lakes with their bases in water. On the one hand, orchids can be large plants, such as the vanilla vines that climb up forest trees and the sobralias, whose canes form large clumps that resemble bamboos. On the other hand, many orchids, especially the epiphytes of smaller branches and twigs, are dwarf plants, often only a few centimeters tall or even smaller. Flowers also vary in magnitude to a remarkable degree. For example, the recently discovered slipper orchid *Phragmipedium kovachii* has rich purple flowers more than 8 inches (20 cm) across, whereas flowers of the tiny epiphytic *Lepanthes calodictyon* are nearly microscopic (2–3 mm across). It is truly remarkable that all these belong to the same plant family, now known to be perhaps the largest family of flowering plants.

Orchids are grouped into a number of genera, assemblages of more or less similar and closely related species that are more closely related to one another than to species in other genera. Many genera are very distinctive, with small numbers of species. Occasionally, a genus may contain numerous morphologically somewhat similar species. The latter often require an expert for definitive identification. Nevertheless, orchid identification is nowadays much easier than it was before because of the number of field guides and illustrated books with fine color photographs that are available. The main challenge is the sheer number of orchid species that occur in the American tropics. *Orchids of Tropical America* provides an easily accessible and finely illustrated beginner's guide to the diversity of genera, the first stop on the way to orchid species identification. Technical language, which abounds in botany and orchid biology, is reduced to a minimum. The fascinating morphology, diversity, ecology, and conservation of orchids are discussed, while some 120 genera of orchids are considered in some detail. The authors conclude with suggestions for some of the best places to see orchids in the wild in the tropical Americas, a great way to start a hobby that is liable, as many have discovered, to become something of an obsession.

The authors of this introductory text are acknowledged experts on tropical American orchids with decades of combined experience of orchids in both the wild and culture. Joe Meisel, based in Ecuador, is an eminent field ecologist whose Ceiba Foundation for Tropical Conservation has been working with local landowners to conserve orchid-rich habitats. Ron Kaufmann, an academic based in San Diego, has a passion for orchid-growing and is a founding member of the Orchid Conservation Alliance. Franco Pupulin, with bases in both Costa Rica and Ecuador, is a professional botanist and one of the most prolific orchid specialists writing on New World orchids. You could not find a better trio of enthusiasts and specialists to introduce you to the wonderful and occasionally bizarre world of tropical American orchids.

PHILLIP J. CRIBB
Royal Botanic Gardens, Kew

Acknowledgments

The authors express their deep gratitude to the generous people who provided photographs, in particular Andy Phillips, Ron Parsons, José Portilla, Fred Clarke, Valérie Léonard, and Alex Zaslawski. Franco Pupulin: I acknowledge the University of Costa Rica for the opportunity to dedicate my time to the project. Ron Kaufmann: I thank all of the people who have spent time in the field with me, sharing their knowledge and love of orchids and the beautiful and often fragile environments in which they grow. The University of San Diego provided invaluable support during the course of this project, and I am grateful for the flexibility that made my participation possible. Joe Meisel: I am grateful for the education, hospitality, and encouragement I have received from numerous orchid and botanical experts, including Monica de Navarro, Harry and Rosemary Zelenko, José Portilla, Ivan Portilla, Hugo Medina, Vlastimil Zak, Peter Tobias, Lou Jost, Alex Hirtz, Rudolf Jenny, and many more. The owners and operators of the El Pahuma Orchid Reserve provided me with my first contact with orchids and the idea of orchid reserves supporting local people; I thank Efrain, Paulina, Rene, Ruth, Roberto, and Marixa Lima, and their sons and daughters, for their dedication and vision. I congratulate, and thank, the following institutions for providing support for orchid research and conservation, to both myself and individuals around the world: American Orchid Society, San Diego County Orchid Society, Orchid Conservation Alliance, Fauna and Flora International, New Hampshire Orchid Society, Madison Orchid Society, Mid-America Orchid Congress, and Greater Kansas City Orchid Society. I acknowledge the hard work of the staff at Cornell University Press, including Katherine Liu, Heidi Lovette, and Peter Potter, and thank them for taking a chance on a new author. The comments of two anonymous reviewers and Candace Akins, the copyeditor, were exceptionally helpful in improving the finished work. It should go without saying that I am grateful to my coauthors Ron Kaufmann and Franco Pupulin whose tireless assistance and superb attention to detail made the collaborative process extremely rewarding. We are all exceptionally grateful for the superb artwork provided by Kathy Creger, who generously contributed her time and creativity for this project. I thank the special collections department of the San Diego State University library for allowing me access to their rare orchid manuscripts.

I was fortunate to obtain an honorary position with the University of Wisconsin's Land Tenure Center & Nelson Institute while conducting some of my research and writing, and I thank Dr. Lisa Naughton for facilitating this fellowship. With deep gratitude I acknowledge the hard work of the staff of the Ceiba Foundation for Tropical Conservation, who put up with a steep decline in my availability during this project; in particular, high praise goes to Carmen Játiva, Emily Lind, and James Madden. Significant contributions to this project were made by Caitlin Murphy, who provided valuable research and assistance. I am privileged to be part of a strong network of intelligent and thought-provoking friends in Wisconsin whose unremitting familial warmth have been invaluable to me: John Robinson, Ritt and Victoria Deitz, Zeke and Michelle Crabb, Andy and Jenny Wallman, Cheryl DeWelt, Didi Guse, and the irrepressible Julie Kreunen. Thanks in particular to Catherine Woodward for inspiring my interest in the amazing world of plants and for accompanying me on a journey to the forests of Ecuador that has lasted more than 15 joyful years. I have been steadfastly supported and encouraged by my family, Joe and Linda and Chris Meisel: many happy summers spent exploring creekbeds near the Mississippi River gave me a curiosity about nature that I carry to this day. Finally, a special word of gratitude to Millionario, a constant companion through interminable hours sitting at a desk.

Orchids of
Tropical America

Introduction

Before the first glow of morning crawls through the dense fog, you wake to the whistles and croaks of hidden birds echoing off the steep mountain slopes. A light breeze carries the sounds of insects, the crackling of wings and scraping of leg-on-leg, along with the musical burble of a rocky stream and the distant roar of a plunging waterfall. The air, still cool from a starry night amid the peaks, delivers a rich texture of fragrances to your nose, exhaled from the damp earth, growing vegetation, and nocturnal flowers. You slip on your boots, step out into the morning, and all around you the forest of a tropical mountain is coming to life.

Your eyes adjust to the diffuse light. The sun will remain hidden behind a distant ridge for more than an hour, but already its rays are scattered through the ubiquitous clouds, giving the eerie sensation of dim illumination from all sides. The canopy overhead remains draped in mist, but all around you the forest reveals itself. Gnarled branches swoop down from thick trunks, bending, twisting, and curling upward again. Ropes and tangles and nets of vines, some as thick as your leg, dangle from the trees, touch the ground, and climb once again. And absolutely everywhere, draped like countless green waterfalls, a thick layer of moss coats all available surfaces. Warm air rising from the lowlands tickles this verdant carpet, causing it to sway lightly as if the entire forest were breathing.

As your gaze wanders over the dense array of vegetation, you see the trunks and branches harbor not only moss but also innumerable plants that seem to sprout directly from the trees: leaves in spiky rosettes, some with brilliant red tips, giant slabs of green projecting atop long stems like umbrellas, cane-like and grass-like ribbons, tiny but succulent leaves, and huge fan-like fronds resembling palms. Amid the profusion of greenery you spy a splash of color. High overhead, a spray of golden flowers erupts from scrambling, twining stems like fireworks frozen in mid-air. Intrigued, you search elsewhere, and are rewarded with more visual explosions: dappled pink to your left, deep vermillion beyond that, a shock of bright purple next to you. And then, directly before your eyes, like a ghost stepping into a spotlight, an enormous white flower materializes, dangling gracefully from a short stalk, complex in form and breathtakingly beautiful. You have just encountered your first wild orchid, waiting patiently for a pollinator to visit its showy bloom, and for the mountain winds to bear its minuscule seeds to new perches in the misty forest.

In the past, such encounters with wild orchids were experienced by only a handful of people. Since their first discovery, orchids have been suffused with an air of romantic mystery, opulent luxury, and entrancing rarity. Stunning flowers of a thousand shapes and hues, collected with great difficulty in the distant tropics, were maddeningly troublesome to cultivate but avidly sought by kings and queens of Europe. Intrepid explorers braved dispiriting weather, disease, and hostile tribes to discover and gather these magical plants. The hunt was on, around the known, and unknown, world.

Like miners pursuing rich veins of gold or prospectors searching for rare gemstones, orchid collectors known as *travelers* scoured the globe for new species. They were given marching orders by dozens of plant nurseries and greenhouses in Europe seeking to capitalize on the ravenous demand for these alluring tropical flowers, in a desperate game of one-upmanship. When a new, showy orchid was discovered in the jungles of South America, provided the plant survived the long voyage across the Atlantic, it could ignite a firestorm of excitement among private collectors on the continent. A single plant fetched obscene prices for the orchid house that owned it, and the pressure to find never-before-seen species was enormous. Some greenhouses had as many as 60 travelers in their employ, fanned out across the mountains and forests of the world's tropical regions. Legends were born, fame was earned, enemies made, and riches won by those few hardy souls who could survive the tropics and bring home the ultimate prize: a flower nobody had ever seen, a gem that everybody wanted.

Rarely has the search for biodiversity on planet Earth been driven so purely by the implacable force of commerce. In modern times, we commonly hear that we should protect the rainforest because it might be home to new cures for cancer, better anesthetics, and more effective vaccines. Potential economic value is inherent in diversity, the argument goes. But in the 18th and 19th centuries, when a novel orchid could attract raving buyers to an English greenhouse, there was a palpable and immediate value to diversity. Newness was the very currency of the trade. And while today the hunt for new species in far-flung and inaccessible regions is recognized as the purview of naturalists and conservationists, in those days the world was explored by the orchid travelers. They paddled uncharted rivers, climbed unmapped peaks, opened new trails, braved thieves and hostile indigenous peoples, and ultimately achieved a nearly complete exploration of the world's tropical regions.

The travelers had a dark side as well. They fought among themselves, occasionally to the death. They sabotaged rival expeditions. But more cruelly, or perhaps out of ignorance and greed, they left a wake of environmental destruction wherever they went. One tree after another was felled in order to harvest the orchids growing in their uppermost branches. If the orchids were within easy reach, entire populations were collected, down to the last plant. Sometimes this wholesale ransacking was done for obvious commercial benefit, but other times for malevolent purposes: simply to ensure that later travelers would be unable to find the species and ship it to a rival orchid house. One traveler, describing a particularly productive site in Colombia, gloated, "I shall clear them all out" (letter from Carl Johannsen to

F. Sander, 1896; reported in Boyle 1901). In this fashion, whole forests were ruthlessly cut down to gather, and eliminate, their orchids.

Back in Europe, as the popularity of orchids grew, so did the intensity of efforts to reproduce them in captivity. Nearly all trials ended in failure, however, increasing the already incandescent pressure on collecting wild specimens. Without the means to reproduce orchids in captivity, the only sources of marketable plants were the remote forests of the tropics, and the travelers.

A sequence of discoveries, some serendipitous, others hard-won in the laboratory, eventually revealed the secrets of orchid reproduction. An invisible fungus proved to be the missing link: without it, orchid seeds could not gather nutrients and failed to germinate. Once the symbiosis with fungus was revealed, commercial production of orchids shifted to climate-controlled greenhouses. The largest, most colorful, vigorous, and long-lived flowers were selected, propagated, and sold by the thousands. Crossbreeding trials yielded eminently marketable hybrids that featured the best traits of each species: some were chosen for the breadth of their flowers, others for the color, still others for their long life or rich fragrance. Today a staggering variety of orchid species and hybrids are produced in colossal greenhouses around the world, in quantities that rival all other potted plants.

And yet that initial urge to experience something new remains strong in orchid lovers. At orchid exhibitions, enthusiasts inevitably crowd around the table that displays a newfound species or novel hybrid, something nobody has seen before. The allure of the unique, the pleasure of expanding one's knowledge, the romance of touching a distant corner of the planet, are powerful forces. This deep-seated urge to expand our horizons leads newcomers to buy their first orchid, hobbyists to install humidifying systems, and experts to build larger greenhouses. The same seductive force entices orchid lovers around the world to join orchid tours to the tropics, where they can see some favorite plants, and many new ones, in pristine native habitat.

Despite the relative ease of international travel, even today most flower lovers catch their first glimpse of an orchid behind the glass case at a neighborhood florist, in their local botanical garden, or in an exhibition dominated by hybrids. Orchid tourism, however, is becoming widespread and within the reach of most enthusiasts. Tropical countries throughout the world boast of their orchid diversity, encouraging tourists to visit the richest regions. Often these are in forested mountains, where small reserves are springing up to protect and showcase local orchid jewels. Orchid tourism, still in its infancy, is providing a novel source of income to local people living in areas of high orchid diversity. Indeed, the revenue derived from orchid tourism may dissuade landowners from cutting forest and converting land to commercial purposes, damaging not only orchid populations but also the incredible diversity of other organisms that thrive in forested mountains: hummingbirds, palms, frogs, ferns, beetles, bromeliads, and many more.

In the past, while driving along mountainous tropical roads, it was disappointingly common to encounter *campesinos* (country folk) in patched clothes patiently holding aloft dazzling orchids in bloom, for sale to those passing by. Such plants invariably had been stripped out of surrounding forests; denied their natural habitat of fresh, humid air and dappled sun, they were consigned to a short life on a kitchen table until succumbing to the stale, dry air of the city. Like the illegal sale of parrots, monkeys, and other wildlife along tropical motorways, these abducted orchids would never be viewed in their native environment, and their intimate and wondrous connections to the forest never fully appreciated. With the rise of orchid tourism, however, such roadside encounters are increasingly rare, and rural people across tropical America have turned to the establishment of modest ecotourism reserves where visitors can see orchids in bloom throughout the year. That same *campesino* might now lead you on a walk amid dozens of flowering orchids, converse knowledgeably about their pollinators and preferred habitat, and extend an invitation to return next month when other species will blossom.

Orchids of Tropical America was written for all lovers of orchids, including those who already have seen a flowering plant in the wild and those who only dream of such a moment. If you are planning a trip to an orchid-rich country in tropical America, this book offers an introduction to these diverse and stunning plants, along with an easy-to-use field guide to more than 120 of their most widespread and species-rich groups. If you are a budding enthusiast with a collection of orchids at home, the book provides detailed information about orchid diversity and ecology, and clear descriptions of orchid groups that are often misidentified. And even if you never visit orchids in their tropical homelands, nor keep these fastidious plants in your home, the pages that follow present a collection of entertaining tales about the fascinating role of orchids in human history.

Orchids are almost certainly the most diverse family of plants on planet Earth, with as many as 25,000 to 35,000 species described and more being discovered each day. Although orchids do occur in the temperate climates of Europe and the United States, the real epicenter of their explosive diversification is found in the tropics. For example, the country of Ecuador, only the size of the state of Colorado, boasts of more than 4000 known species, more than one-tenth of all described orchid species in the world. Nobody knows the full explanation for the immense diversity of these odd-looking plants with hypnotically beautiful flowers, since no human bore witness to the explosion of most of the present-day species. We do understand, however, some of the evolutionary pressures that contribute to the enormous array of colors and forms, and many of these processes are actively at work even today.

From blooms larger than your hand to some so tiny they can be appreciated only with a microscope, orchids present an astonishing array of flower shapes, sizes, and colors. Some are highly decorated, bursting with bright colors, impressionistic flecks, and mesmerizing patterns. Others are drab, green, and inconspicuous. While most orchids lack the pleasant fragrance we usually associate with attractive blossoms, many species do produce strong and distinctive odors. Some exude a delicious perfume, particularly those that attract scent-sensitive moths for nocturnal pollination. Others produce an aroma that, while offensive to the human nose, is designed to seduce flies sniffing the air for signs of rotting fruit or even dead animals. Scented or odorless, flowers can be borne in great numbers on long stalks, high above the plant or hanging far below; some stalks can reach 10 feet (3 m) in length or more. Other flowers are held close to the plant, appearing in some species to be glued to the upper surface of the leaf. Some orchids explode into bloom with dozens or hundreds of flowers at once, while other species unveil only a single flower at a time, which if pollinated induces the termination of all remaining flower buds.

Apart from the multiplicity of flower types, the plants themselves exhibit an enormous range of forms. Grass-like orchids grow in carpets on boulders, cane-like plants cover open ground in dense thickets, while the majority of the species are found growing in trees as epiphytes (literally "on plants"). Leaves can be thick, shiny, and leathery like a new belt, succulent like an aloe, or bristling with dense hairs. Some sprout only a whorl of broad leaves, lying flat on the soil surface. Other fronds are delicate, thin, and wrinkled, or long and narrow resembling the leaves of a lily. Typical orchids are relatively small and bear few leaves, but some can develop into bountiful shrubs with robust vegetative growth. Some extend laterally, with successive tufts of leaves sprouting like consecutive ranks in a marching band. Others climb skyward, adding new growth only to the plant's tip, like segments on a tent pole. Roots, hidden in ground-dwelling orchids, reveal themselves in epiphytes as a twining, twisting mat of thick, white, and waxy fingers clasping their arboreal perch. Many orchids that are adapted to life in the trees face regular shortages of water and utilize a strategy perfected by desert camels: they store water in their own humps, swollen green *pseudobulbs* (see Illustrated Glossary for detailed definition), found between the roots and stems, usually thickened into shapes resembling a vase, spindle, or football. This range of forms, like a deck of a thousand cards, offers orchids virtually endless permutations of flower, leaf, and stem. The subsequent diversity of structures and adaptations has permitted orchids to colonize nearly all of planet Earth.

Although it is widely recognized that orchids reach their peak abundance and variety in the mountainous forests of the tropics, they are such a diverse group that representatives can be found almost anywhere in the world. They appear on all the continents except Antarctica and occur in virtually every habitat type except glaciers and, of course, the ocean. Orchids grow at near-freezing temperatures above 13,000 feet (4000 m) on some equatorial mountains, in dry forests dominated by cactus and thorny shrubs, in open grasslands more suited to antelope and buffalo, and in the cool forests of the temperate latitudes. Despite their reputation for requiring a warm and wet environment, orchids grow in the chilly climes of Canada, Siberia, and the Himalayas, where only terrestrial species occur. Other species survive the scorching temperatures of the desert: amazingly, several Australian species grow and flower entirely underground to conserve water in the blazing heat of the outback.

Why are orchids so diverse? Many reasons for the rapid explosion in the number of orchid species are known, although we still lack a comprehensive explanation of why these unique plants achieved such immense diversity. Some of the primary causes include the isolation of small populations of orchids by the soaring ridges and deep valleys of tropical mountains, and the fantastically intricate relationships between orchid flowers and their pollinators. We will discuss the diversity of orchids in greater detail in a subsequent chapter, but the end result is simple enough to comprehend: orchids make up around one-tenth of all plant species on Earth, and exceed, by themselves, the total plant diversity of North America.

Despite these daunting numbers we should not be discouraged from seeking to learn more about orchids and to recognize their major groups. Although the exact number of species is huge, and climbing, all orchids are sorted into a manageable number of groups, each known scientifically as a genus (plural: genera). These groups can be identified by a relatively small number of prominent, distinctive physical features. Once you learn to recognize a handful of these visible characteristics, you can place almost any orchid into its proper genus. Certainly, most of us are able to distinguish a rose from a tulip, or a pansy from a daisy, even if we do not know exactly what separates these different flowers. It is a matter of observing plants in slightly greater detail, recognizing certain important traits, and having an informative guide that describes genera in terms of these characteristics. That guide is precisely what you hold in your hand: *Orchids of Tropical America* will help you recognize key physical attributes of orchids, then use combinations of those attributes to identify most orchids to genus. The book provides photographs, detailed descriptions, and extensive ecological and historical information about the most widespread and common groups of tropical orchids from Mexico to Brazil.

The authors share a passion not only for the beauty and diversity of orchids but also for the protection of the varied habitats where they are found. In particular, high-elevation mountain forests where orchids reach their peak abundance and diversity are exceedingly important, not only to these magnificent flowers but also to countless other plant and animal species. Towering trees and the dense mosaic of plants that drape them, endangered mammals such as the Andean spectacled bear and mountain tapir, hummingbirds that buzz through the canopy like flying jewels, frogs that fill the night with a symphony of trills—all these species call the mountain slopes home. Indeed, for the same reason that orchids are so diverse in rugged mountain ranges, these other plants and animals also have evolved into many species, often separated by only short distances. This phenomenon is called *endemism*, when distinct species have very small geographic ranges. Mountainous tropical forests have some of the highest rates of endemism on Earth, and therefore protect some of the greatest densities of unique species.

Steep slopes cloaked with forest also provide an indispensable ecological service to the towns and villages located downslope from them. Vapor from the clouds that envelop such mountains nearly every day condenses on foliage, providing moisture to the thick vegetation. This water eventually makes its way into the numerous streams and rivers that flow out of the forests. In fact, mountain forests and other high-elevation habitats produce the majority of drinking and irrigation water used by people throughout the tropics. Where these forests have been cut, streams inevitably run shallow or dry up altogether, to the detriment of farms, reservoirs, and towns below. Even worse, when steep hillsides lose their thick covering of vegetation, roots no longer hold the soil in place and catastrophic landslides become commonplace. Thus, conserving intact habitat in the tropical mountains protects enormous numbers of unique species and preserves a way of life for the people living in the region.

For these reasons, we salute projects throughout the tropics dedicated to the conservation of pristine orchid habitat. We provide a section in this guide that offers information about reserves and parks across tropical America that protect these forests and provide visitors an opportunity to see firsthand the orchids and other species that dwell within. Our wish is that this book will expose to you the beauty and diversity of orchids, and encourage you to support protection of these fantastic plants and the habitats in which they thrive. Deforestation, theft of plants, uncontrolled fires, wildlife poaching, and other activities continue to threaten the integrity of tropical forests, but successful conservation projects around the globe demonstrate just how much can be achieved when informed and energized people make a priority of protecting wild places. Even if you never awake on a misty morning in a mountainous forest to discover an orchid for yourself, you can do much to ensure that those orchids, and those misty mountains, continue to provide ecological services and beauty to residents and visitors around the world.

Orchid Ecology, Diversity, and Conservation

Orchids have been a source of amazement, amusement, and scientific discovery for more than 500 years. Early explorers and collectors were captivated by the shapes, colors, and species diversity of this beguiling family of plants. Later researchers, such as Joseph Hooker and Charles Darwin, spent countless hours studying the intricacies of flower development, pollination, and germination. Pioneering orchid expert John Lindley eloquently summarized the allure of orchids: "Whether we consider general elegance of individuals, durability of blossoms, splendid colours, delicious perfume, or extraordinary structure, it would be difficult to select any [family] superior to Orchideae in these respects, and few even equal to them" (Ossenbach 2009: 62).

Fascination with the study of orchids has continued into the modern era. Each year, hundreds of scientific studies are published that illuminate new aspects of their intricate and remarkable lives. While an exhaustive treatment of the ecology and biology of this multitudinous family is beyond the scope of this guide, a brief introduction to the complex web of interactions that knit orchids to their environment will help inspire a richer understanding of these mysterious and majestic plants.

Biology and Ecology

Biology is the study of living organisms, how they grow and reproduce, and the function of their bodies. *Ecology* focuses on the interactions between an organism and its surrounding environment, including other organisms with which it is linked. In this book, we weave the two topics together, offering ecological stories that shed light on the biology of plants in each genus. To better understand and appreciate these stories, we offer here a primer on the biology and ecology of orchids that provides fundamental background information and introduces a few key terms.

Seeds and Dispersal

Let us start at the beginning of an orchid's life cycle. Floating on a faint breeze, orchid seeds are borne gently through the air, eventually coming into contact with a landing site suitable for their growth. These seeds are astonishingly tiny, among the smallest in the plant kingdom, and more than a million can be packed into one diminutive fruit. A single seed can be less than 1/150th of an inch across (0.2 mm), about three times the diameter of a human hair, and tip the scales at just 40 millionths of an ounce, or one-tenth the weight of a grain of sand. Why are orchid seeds so minuscule, and how can they be so tiny yet still grow into a large and vigorous plant?

Although answering "why" questions in biology can be difficult, the prevailing theory is that orchids produce huge numbers of seeds to ensure that at least some will alight on an appropriate substrate, a location that meets the plant's strict requirements for germination and growth. If only a few seeds were released they might all fall onto poor-quality sites, perhaps too soggy, too scorching, or too rocky, and the plant would fail to reproduce. Virtually all orchids rely on wind to broadcast their seeds, and small, lightweight seeds are more likely to float farther on the slightest breeze and encounter a more diverse range of landing sites, at least some of which will be acceptable.

Because each orchid species has such stringent requirements for survival and growth (ask any greenhouse manager just how finicky they can be), most of the seeds will fail to germinate. Only a tiny fraction will beat the odds and randomly strike an acceptable patch of soil, bark, rock, or moss on which to settle and grow. Indeed, if each and every seed released were to successfully germinate, the rate of orchid population growth would be astronomical. Darwin estimated "the great grandchildren of a single plant would nearly . . . clothe with one uniform green carpet the entire surface of the land throughout the globe" (Darwin 1877: 277). That an orchid species does not achieve such global carpeting attests to the rarity of perfectly favorable sites, and the need to disperse mote-like seeds as widely as possible.

How orchid seeds can be so small is simpler to explain. They achieve their tiny size because they lack the weighty tissue that usually nourishes the seed, the endosperm. When you eat a peanut, pecan, or walnut, you are eating the endosperm. This nutrient-rich nutmeat, in most plants, feeds the developing embryo as it grows from seed to seedling. Without a bulky endosperm the orchid "seed," more properly called an embryo, is incredibly light in weight and can travel in the air for miles on the faintest whiff of wind. A few studies have recorded astonishing seed dispersal distances of 600 to 1200 miles (1000–2000 km).

An additional benefit of the lack of endosperm has been suggested that may explain the sprawling diversity of the family. In many higher plants, endosperm actually blocks the growth of hybrid seeds, those produced by cross-pollination between species. The endosperm serves as a kind of gatekeeper, preventing unwanted hybrids from creeping into the population. Hybridization, however, has fueled the explosion in orchid diversity and has abetted their successful colonization of extremely varied habitats. The lack of endosperm may facilitate the sharing of genes between species, permitting accelerated adaptation to local conditions that has proved beneficial to orchids (albeit occasionally frustrating to taxonomists).

Germination and Symbiotic Fungus

Orchids pay a price for the absence of endosperm because their seeds have no stored nutrients to assist germination and growth. Instead, virtually all orchids rely on invisible fungi in the soil, moss, bark, or other substrate where the seeds land. The fungus swiftly invades the seed and begins transferring nutrients and minerals from the substrate to the germinating embryo by digesting organic matter in the environment and converting it to forms more readily absorbed by the young plant. The two organisms function together, in symbiosis, but only if the type of fungus is a perfect match for the species of orchid. This one-to-one matching is of critical importance and was the cause of countless failures in commercial orchid breeding before the role of the fungus was elucidated.

Upon germination, most orchids develop tiny embryonic leaves that begin the process of photosynthesis and roots that grasp the substrate. The fungus eventually takes up residence in the roots, where it is known technically as a *mycorrhiza* (derives from the Greek for "root fungus"). Over the course of the orchid's life, the two organisms enjoy a mutually beneficial coexistence: the plant provides the fungus with sugar and starch, products of photosynthesis, while the fungus continues to assist the roots with uptake of necessary minerals.

A few species of orchid (e.g., *Govenia*) postpone the development of leaves, however, living entirely underground for years while feeding on decaying organic matter. In this they are similar to traditional fungus, which consumes soil materials to gain energy before producing mushrooms. Once conditions are ideal and sufficient nutrients have been amassed by the subterranean orchid, a process that can take years, the plant emerges into the light, yields an eruption of flowers, and scatters seeds to found the next generation.

In addition to encountering the proper strain of fungus, all orchid seeds must find an exact mix of environmental conditions. These include biotic factors such as bark type or leaf litter composition, and abiotic (i.e., nonliving) factors such as elevation, temperature, light intensity, and moisture levels. Orchids may fail to germinate or flower because of only the slightest variation in some of these conditions. Matching a species to its particular, intricate blend of factors is one of the greatest challenges to botanical gardens and greenhouses, from small household collections to mammoth commercial operations.

Nutrients and Water

Once leaves have emerged, photosynthesis can begin. Like tiny solar panels, leaf cells capture light energy, then use that energy to convert atmospheric carbon dioxide into starches and sugars, the building blocks of life on Earth. But all living organisms, plants included, need much more than just starch. They require amino acids to build proteins and DNA, fats to construct cell membranes, and minerals including calcium, phosphorus, nitrogen, and magnesium that are critical to the underlying chemical processes of life.

Most of those key minerals are water soluble and can be drawn up by the roots along with water held in damp moss, muddy soil, or wet bark. The plant uses a mechanism called *transpiration* to pull water into the roots. Moisture evaporates constantly from tiny pores in the leaves, called *stomata*, creating a slight negative pressure that draws water from the soil into the roots, much like sucking on the top of a straw draws liquid into the bottom. The minerals then travel up the orchid, pulled in a rising column of mineral-rich water through microscopic plumbing in the stem.

Because of this dependency on moisture, orchids, and epiphytic orchids in particular, face powerful water challenges. Life in the treetops, perched on a branch or clinging to the trunk, provides benefits such as ample sunlight and wind for seed dispersal but severs the connection between the plant and a reliable reservoir of moisture, the soil. Shortages of available water are commonplace. Even a few hours without rain or clouds can lead to a treetop perch desiccating almost completely. These epiphytes also lack access to the nutrient storehouse the soil provides, a difficulty that rarely affects terrestrial plants. To withstand the scarcity of nutrients and water, epiphytic and lithophytic (rock-dwelling) orchids have adopted a variety of strategies and structures.

Orchids in the tree canopy frequently colonize sites in which small patches of humus accumulate. Fallen leaves, twigs, and other vegetative material collect in shallow depressions where branches connect to trunks, in small cavities in a bole or branch, and in the furrows of rough bark. This material decomposes over time, forming a sort of canopy soil that epiphytes can utilize, and they extend fine roots into the soil patch to access its nutrients. Epiphytes can even form their own soil by capturing falling leaves and holding them near: some orchids (e.g., *Gongora* and *Mormodes*) have upturned roots or protruding spines that catch organic material, known as *throughfall*, like a basket. The leaves decompose within the basket, and the plant reaps the benefits. These so-called trash roots are, incidentally, not uncommon in terrestrial plants that grow in poor soils and face similar nutrient shortages.

Such strategies provide epiphytes with access to modest plots of soil that soak up moderate amounts of rainfall and hold the moisture for a few days before drying out. Where rainfall is more regular, and water shortages infrequent, a greater abundance and diversity of orchids can prosper. For this reason, tropical montane forests, known colloquially as cloud forests, harbor the highest diversity and density of orchids in the world. Their astounding abundance is further promoted by the ubiquitous light that penetrates uneven tree canopies jostling on steep slopes, so different from the deep shade found beneath a tall rainforest. Cloud forest trees are positively festooned with epiphytic plants: one famous survey in Venezuela found 47 species growing on a single tree.

Once canopy orchids acquire precious moisture, they must protect it at all costs, and several structures exhibit modifications for storing water. Leaves are thick and succulent, or juicy. Roots are covered in a whitish, waxy material called *velamen* that can absorb water rapidly from the surrounding environment like a sponge, and slow its evaporation from within the root, like plastic wrap on a bowl of leftovers. Water also is stored in *pseudobulbs*, the curiously swollen stems so characteristic of epiphytic orchids (and largely absent from terrestrial species). Ranging in shape from squat onions to elongate cylinders, the pseudobulbs swell tightly when moisture is abundant but lose their plumpness, becoming partly shriveled, in times of water scarcity. Pseudobulbs further serve as warehouses for collected nutrients, holding them until they are mobilized for use by the plant during dry, challenging times.

The first foreign explorers of the New World referred erroneously to epiphytic orchids as parasitic air plants, presuming that they fed on the sap of their host trees (like mistletoe). Orchids are not parasites, however, and use trees only for structural support and for the meager accumulations of soil encountered upon them. This misapprehension is not surprising, as epiphytes have few visible sources of nutrients, severed as they are from the soil. Some nutrients may be delivered almost invisibly through dry deposition, falling from the sky as dust; this route is the most common source of phosphorus, for example. Dust thinly coats the orchid's leaves and stem, and later is rinsed toward the roots by intermittent rains. Water dripping through the canopy also washes accumulated dust and minerals off leaves overhead (of the tree itself and numerous other epiphytes) and onto orchids below. The water, now enriched, reaches the orchids by falling directly on them as rain (the previously mentioned throughfall) or by coursing down the branches and trunk, a pathway known as *stemflow*.

Many of these adaptations can be observed not only in epiphytic orchids but also in terrestrial plants growing in nutrient-poor soils and in arid climates. The water stress faced by most canopy epiphytes is also a challenge to orchids growing in deserts, or on rock faces where sheer expanses of stone retain little or no moisture. Thus one can observe numerous similarities between dry climate orchids and epiphytes: both evince adaptations to collecting scarce water and storing it through periods of drought. Orchids that grow on poor soils, such as the highly leached clays of the Amazon basin, face significant nutrient limitations and also must rely on throughfall and stemflow to survive.

Growth Forms

Let us return to the growth of the newly germinated plant, its fungus-assisted infancy now safely in the past. Photosynthesis in the young leaves, coupled with nutrient and water resources, fuels the development of new stem and leaf tissue. This growth may arise from atop a pseudobulb, if present, or sprout directly from the roots. Gradually, the plant acquires its own characteristic shape. Myriad forms exist in the hyper-diverse world of orchids, from tiny moss-like plantlets to scrambling tangles to hefty shrubs and even to vines more than 40 feet (12 m) long. But most orchids share an aspiration to height, thrusting stems and leaves upward to gather light and to lift flowers higher into the air.

As vertical development proceeds, many orchids also begin to show lateral growth. Horizontal runners called *rhizomes* emerge from the base of the stem or pseudobulb, traveling on top of or just beneath the soil surface. From the rhizome, new pseudobulbs or stems eventually emerge. During this process, in pseudobulbous species, leaves are shed from the older pseudobulb, which is now referred to as the *back bulb*. This aging pseudobulb begins to senesce after some time, turning brown and shriveled before decaying entirely. Novice growers occasionally mistake this decline as a sign of illness. The orchid is merely mobilizing the back bulb's nutrients, however, to support growth of the new pseudobulb, called the *front bulb*.

In most Old World orchids (those from Europe, Asia, Africa, and Australia), rhizomes are absent and new tissues are continually added atop the previous leaves and stems, a growth form called *monopodial* (e.g., *Vanda*). New World orchids (from North, Central, and South America, and the Caribbean), for the most part, follow a *sympodial* growth form characterized by lateral rhizomes and multiple, parallel stems or pseudobulbs (e.g., *Cattleya*). As rhizomes spread, the orchid can become quite mat-like, and in some instances a single plant may cover dozens of square feet with scrambling stems and runners.

Defenses and Defenders

While the young orchid is growing and developing, it faces an onslaught of pests and diseases. Plants in the tropics, where no harsh winters beat back insect populations, are particularly threatened by herbivores, fungus, and disease. For most plants, battling these attackers is a lifelong struggle. Orchids utilize a variety of techniques to reduce the maleficent effect of pests, strategies shared by other kinds of plants. Leaves often have a thick, waxy covering, or cuticle, that cannot be chewed by the weak mandibles of small insects. In other species, leaves and stems may be covered with dense hairs (e.g., *Dresslerella*), a deterrent against insects that find it difficult to walk on the layer of prickly fuzz, much less burrow through it to reach juicy leaf tissues. Still others impregnate leaves with harsh chemical compounds distasteful to would-be herbivores (e.g., *Habenaria*).

A few groups of orchids have adopted a more mercenary strategy for defense: independent contractors. These orchids have established tight relationships with ant colonies, which aggressively defend the plants against insect pests such as beetles, caterpillars, and aphids. In exchange, the orchids offer homes for the ants in hollow pseudobulbs (e.g., *Myrmecophila*) or provide sugary rewards from glands mounted on stems, buds, and flowers (e.g., *Cycnoches* and others). The ants and orchids live together in mutual benefit, a situation also observed in non-orchid plants such as *Acacia* trees. Other orchids prefer greater firepower, favoring sites occupied by wasp nests, from which emerge vigorously stinging defenders when the branch is shaken or the plant disturbed.

Flowers and Flowering

If the early years of growth are survived, and pests and diseases are successfully battled, a plant may finally reach reproductive age and yield the spectacular flowers that have earned orchids their worldwide fame. Flowers evolved to attract pollinators, which carry genetic material between blossoms of widely separated plants. The resulting mixture leads to nearly infinite combinations of DNA in the subsequent generation of plants, of which only the hardiest typically survive to reproduce. The endless diversity of forms permitted by reproduction through pollination gives the species the greatest chance of survival. Natural selection reinforces the most successful traits, winnows out genetic failures, and the population becomes better adapted to local conditions over successive generations. Thus flowers offer tremendous evolutionary benefits and orchids invest heavily in them.

Flowers, and the fruits they produce, are expensive to manufacture. Blossoms are complex creations, colorful and commonly fragrant, but temporary: eventually they are shed, falling to the ground and taking with them precious nutrients invested in their construction. Orchids may require several years to accumulate sufficient nutrient stores to power the energy-intensive process of reproduction. Limited water or nutrient availability, climatic extremes, or low light levels can slow the pace of an orchid's maturation. Typically four years or more elapse before a given orchid reaches the age of flowering. In some, this process can be accelerated, as in twig orchids (e.g., *Erycina*) that bloom within one or two years, probably because of the inherent uncertainty of life on thin branches. In others, it can be delayed by the challenges of a harsh environment, insect pests, or a poor growing site.

The exact timing of flowering, and the cues that trigger it, are a mystery in all but a few well-studied species of orchids. Some orchids flower at night, when moths are active. Others bloom in the early morning, or the heat of midday, to coincide with the peak activity of their preferred pollinators. Typically orchids will flower during drier periods of the year, rather than in rainy seasons when heavy downpours ground flying insects and can accelerate the decomposition of blossoms. Flowering usually is timed so that seed capsules mature and release seeds just before the rains begin (in regions with distinct seasons), so that young plants have sufficient moisture for germination and early growth.

To maximize the interchange of genetic material between plants, there has been strong evolutionary pressure on orchids to adopt synchronous flowering. Pollinators visiting the flower of a synchronized orchid are more likely to find other flowers open at that time, enhancing the likelihood that the pollen they carry will reach a receptive flower of the same species. In some areas, entire hillsides can erupt in blossoms of a single orchid species; this spectacle is a delight to orchid enthusiasts, but often fades swiftly. Usually these synchronous flowerings are staggered between different species (e.g., *Sobralia*), with flowers of only one species open at a given time. This offset in timing reduces the likelihood that pollen will be wasted on flowers of an unrelated orchid.

Evolution has worked intensely on the duration of flowering periods as well, by balancing countervailing pressures. The longer a flower is open and receptive, the greater the chance of pollination. Flowers cost energy to maintain, however, and thus represent a drain on the plant's resources. On the one hand, excessively long flowering periods could weaken the plant, undermining its defensive efforts against pests, for example, and potentially lead to its death. On the other hand, flowers that open too briefly risk being overlooked by pollinators and failing to achieve their critical function of reproduction. Thus, many orchids have adopted a diversity of strategies to enhance the attractiveness of floral displays while holding energy costs to a minimum.

Some orchids have large bunches of flowers on a single stalk (called an *inflorescence*) in which individual flowers are open for only a few days each, but the overall display is vividly appealing over the course of weeks or even months.

Others stock the inflorescence with numerous infertile flowers, visually attractive but cheap to produce, amid a smaller number of fertile blossoms. Insects are drawn to the large, flamboyantly colorful display and eventually find the fertile flowers after a bit of trial and error. Bright colors, striking patterns, unusual forms, and strong fragrances round out the set of tools orchids rely on to attract pollinators (see next section).

Plants also can reproduce vegetatively, without depending on flowers and pollinators. Some orchids occasionally produce a thin stalk, often from the inflorescence, that bears a tiny but fully formed orchid called a *keiki* (Hawaiian for baby, pronounced "kay-kee"). The keiki looks like a baby plant on a leash, often dangling in the sky beneath the parent's perch until growth and winds bring it into contact with a suitable substrate. There the keiki, which already has well-developed roots, latches onto the substrate and begins an independent life; the stalk connecting the two quickly disintegrates. In this way, some orchids (e.g., *Epidendrum*) can reproduce asexually; however, the new plant is a clone, an exact genetic copy of the parent. Thus, vegetative reproduction is less beneficial than pollination, which introduces fresh genes into offspring.

Pollinators and Pollination

Insects were the first animal pollinators on earth, and their collaboration in plant reproduction stimulated the appearance of innumerable shapes, colors, and sizes of flowers. Put succinctly, without insects there would be no flowers. The beauty that today entices humans to collect, photograph, and pen verses about flowers exists because flowers evolved to be attractive to insects. Once insect pollination became established as the dominant force in plant reproduction, flowers diversified enormously to match the panoply of pollinators; in turn, booming plant diversity encouraged the appearance of a vast array of new insect species, in a tandem explosion of biodiversity. The profound effect on orchids of this simultaneous diversification inspired a lifetime of passionate study by the great naturalist Charles Darwin: "In my examinations of orchids, hardly any fact has struck me so much as the endless diversities of structure—the prodigality of resources—for gaining the very same end, namely the fertilization of one flower by the pollen from another plant" (Darwin 1877: 284).

The majority of orchid flowers in tropical America are pollinated by small- and medium-sized bees known collectively as *euglossines*, or orchid bees. These industrious insects utilize fragrance chemicals collected from orchids as an integral part of their own courtship routines. Male bees collect these odoriferous compounds from orchid flowers, store them, and later repackage them into mating pheromones that attract females. Orchid blossoms relying on euglossine bees for pollination typically entice them with a combination of brightly patterned colors and a horizontal landing platform. Having alighted and crawled inside the flower, the bees scrabble at various parts of the lip where concealed glands release diverse scents. Often, highly particular blends of distinct odor chemicals are produced from these glands, each unique cocktail appealing to a single species of bee. Such specificity enhances pollination efficiency: the discerning bee will visit only one species of orchid, rather than wasting pollen on flowers of a different species that emit a different fragrance.

Although numerous chemical and mechanical barriers exist, gene flow between separate species does occur in orchids, a process known as *hybridization*. Hybrid orchids are not uncommon, and their existence in nature is considered evidence of ongoing, fast-paced evolution in the family. In some epiphytic orchids from the Andes, a so-called swarm of hybrids can be observed in which multiple putative species share DNA among them, carried by mutual pollinators. Such hybrid swarms are thought to provide these epiphytes with greater advantages when exploiting the highly diverse and mercurial habitats of mountain forest canopies (e.g., *Cycnoches*). Intriguingly, terrestrial orchids that inhabit a more stable environment rarely if ever show high rates of hybridization.

A broad spectrum of insects renders pollination services to orchids. Many orchids are pollinated by bees outside the euglossine group, ranging from small sweat bees to robust carpenter bees. Typically these insects are attracted by nectar reservoirs that the flower provides. Butterfly-pollinated orchids also offer nectar, as do flowers visited by hummingbirds, normally in long tubes called *nectar spurs*. Butterflies tend to prefer brightly colored flowers, typically pink or purple (e.g., *Epidendrum*). Moths are attracted by nocturnal fragrance rather than color, and their flowers tend strongly toward pale green and white colors. Hummingbirds are among the few pollinators capable of seeing red, and the flowers they service are dominated by bright crimson and orange (e.g., *Elleanthus*).

Unusual variations on the theme of attracting insects abound. Some orchids rely on foul odors and purplish or brownish flowers reminiscent of rotting meat to attract flesh flies, the maggots of which must be reared on carrion (e.g., *Bulbophyllum*). Others mimic tiny mushrooms on which minute fungus gnats normally lay their eggs and are even found near clusters of real mushrooms that attract the gnats (e.g., *Dracula*). Orchids even imitate insects themselves, inviting attack or mating attempts by rivals or suitors, respectively (e.g., *Trichoceros*). A few species take advantage of voracious predatory insects by presenting counterfeit prey—spots that look like aphids or bristles resembling spiders—which are mistakenly attacked (e.g., *Brassia*).

In each of these orchid groups, the exact architecture of the flower is critically important. Small flowers bar large insects from entering. Flowers without landing platforms do not attract butterflies, which prefer to alight before feeding on nectar. Tubular flowers are formed in ways that guide the beak of a hummingbird or tongue of a moth down the tunnel to the nectar reservoir they seek. And most important, the presence of diverse creases, humps, wings, and lobes on the flower's lip and column ensure that visiting insects tour the flower in a highly orchestrated manner that ensures they will contact the blossom's reproductive organs in a programmed sequence to successfully effect fertilization.

Insect pollinators, while seeking their reward or departing the flower, are forced to contact the flower's column, which contains the orchid's male and female reproductive organs. Visitors initially pick up pollen, in tiny packets called *pollinia*, which are glued to specific parts of their body: athwart the back, on top of the head, on the face, or on the underside of the abdomen. A pollinium may comprise more than a million pollen grains, enough to produce an equal number of seeds in a single fertilization event. Pollinia usually are mounted on a narrow stalk, the *stipe*, itself attached to a sticky disc. The disc provides the glue that adheres pollinia to pollinator; the stipe ensures precise orientation of pollinia. Flower visitors may instigate the release of pollinia by direct contact, or by brushing sensitive triggers that fire pollinia forward like bolts from a crossbow (e.g., *Catasetum*).

Once laden with their genetic cargo, pollinators depart in search of a second flower. While in flight, the stipe changes shape, straightening or bending as it dries. Once this realignment is complete, a process that requires up to an hour, the pollinia are held in the perfect position to accomplish fertilization. When the pollinator enters the second flower, pollinia held just so are pressed into the stigmatic area of the column, the female reproductive zone. In some orchids, the pollinia are snagged by outcroppings of the column, like a boat hook grasping a deck cleat.

Certain genera of orchids have separate male and female flowers (e.g., *Cycnoches* and *Mormodes*), sometimes on completely distinct plants, although this is rare in the orchid family. The benefit is that self-fertilization in these dioecious (i.e., two houses) species becomes nearly impossible: a bee toting pollinia from male flowers must visit a separate plant, with female flowers, before pollination can be completed. Other orchids bear both flower genders on the same plant (i.e., monoecious), but their maturation is staggered so that male flowers open and close before female flowers are receptive.

The obvious benefits of cross-fertilization have led a small number of orchids to adopt extreme measures to prevent self-pollination. In these flowers, the identity of the incoming pollen is biochemically assessed, and if it was borne from the same plant, the flower releases powerful enzymes that dissolve the reproductive apparatus into mush (e.g., *Coryanthes*). Thus, in these plants, self-pollination is utterly impossible, a strategy that forces the orchid to depend completely on its insect helpers for reproduction.

Because of the unreliability of insect behavior and availability, however, most orchids are at least capable of self-fertilization. Heavy rains, high winds, or the failure of food plants all can cause local pollinator populations to decline, leaving orchids little choice but to engage in self-pollination or risk losing a reproductive season. In such cases, the untouched pollinia dissolve of their own accord and slide, sometimes assisted by rainwater, onto the stigmatic area. There the plant's pollen fertilizes its own eggs, producing offspring that are a re-blending of the single parent's genes, but incorporate no fresh genetic material.

Once pollinia have contacted the stigmatic zone, with or without the aid of a pollinator, the process of fertilization unfolds, albeit largely out of view. The pollen grains packed into the pollinia separate, and long tubes grow from them, through the column tissue. These tubes extend, like tunnels bored through a mountain, until they reach the flower's ovary at the base of the column. Within the ovary lie as many as a million microscopic eggs, and each pollen tube winds its way toward one of them. Once growth of the tubes is complete, mature sperm swim from the pollen grains down the tunnels, reach the eggs at the other end, and accomplish fertilization.

This process is another of the great inventions of the orchid family: a single pollination visit, by a lone insect, can transport enough genetic material in the form of one pollinia to fertilize hundreds of thousands of eggs.

Upon fertilization the ovary begins to swell at the same time that other flower parts senesce. The petals and sepals dry up and shrivel, while the eggs inside the ovary develop. The walls of the ovary thicken and harden and eventually form a *capsule*: a hollow structure containing the fertilized eggs, now tiny embryos. After this capsule matures, a process requiring months to as much as a year, it dries and splits open along several seams, revealing the dry, dust-like seeds inside. With a little luck, and a decent breeze, those seeds will be wafted to a suitable site for germination, where another generation of orchids will embark on a new life.

Diversity and Distribution

Somewhere in western Australia, away from the scorching temperatures of the outback, buried below ground, leafless but not lifeless, a tiny plant bides its time, robbing nutrients from soil fungi and waiting for the precise set of conditions that will trigger it to flower. This subterranean plant is an orchid, and its unconventional life reveals just how fantastically diverse is the family Orchidaceae. When hearing the word *orchid*, most people imagine a wiry stalk arising from a few strap-shaped green leaves to display luxuriously colored, saucer-sized flowers. Although delightful to behold, the plant just described resembles only a fraction of wild orchids. In truth, flowers vary from huge to nearly microscopic, and from vivid to colorless; leaves may be as long as your arm, as small as a dragonfly's wing, or completely absent; some species climb to 40 feet in the canopy, while others go unnoticed without a magnifying glass. Spanning these extremes is the diversity of orchids, one of the most species-rich plant families on Earth.

Estimates vary but place the total number of orchid species between 25,000 and 30,000. These represent approximately 10% of all known flowering plants on the planet. For comparison, there are some 4600 types of mammals in the world, 9000 species of birds, and approximately 11,000 reptiles and amphibians. Thus it takes all four of the most well-known animal groups to equal the number of orchid species.

Given such staggering levels of diversity, it is not surprising that orchids are found in virtually all types of habitat, and on every continent (save Antarctica). Orchids display adaptations to nearly every climate: some species can withstand extreme cold, others blazing heat, some are subterranean, and a few are wholly aquatic. Many orchids are terrestrial, living in forests, bogs, and grasslands, others cling to bare rock, while a large proportion have adopted a life in the trees. Bees pollinate most flowers, but a surprising number rely on hummingbirds, butterflies, moths, and even flies. The exceptional diversity of the family implies that for nearly every lifestyle and set of environmental conditions, there is an orchid that will thrive.

Although orchids occupy most of the globe, their distribution is by no means uniform. A pattern observed in many plant groups also applies to orchids: diversity increases as you approach the equator. For example, a typical forest in North

America supports 30 tree species per hectare (about the size of two football fields); a similarly sized area in the Amazon hosts more than 300. The same pattern applies to birds, mammals, and insects: only 10 species of ants were found in surveys at 60 degrees north latitude (the line connecting Anchorage and Oslo), while similar studies in the equatorial Amazon yielded more than 6000. Orchids fit the trend closely, with around 250 species in the United States (half of them in Florida), but more than 3000 in Colombia and more than 4000 (more than a tenth of the world's diversity) in Ecuador, a country the size of Colorado.

While orchids conform to latitudinal trends, their distribution tends to buck patterns that relate diversity and elevation. Most plant and animal groups reach their greatest diversity at low elevations, and species number drops with increasing altitude. Insects, trees, amphibians, and more are testament to this rule. But orchids diverge. Although many orchid species do occur in lowland tropical rainforests, their greatest diversity is found in middle-elevation mountains, between about 3000 and 6500 feet (900–2000 m) of altitude. In the steep, misty forests on these slopes, thousands of individual plants drape on tree trunks and branches, cling to twigs, and constellate the mossy understory. A single tree may host nearly 50 distinct species of orchids; areas no larger than a bedroom, more than 100.

Where the patterns of altitude and latitude intersect flourishes the greatest diversity of orchids on Earth: the Andes Mountains. Stretching like an off-center spine down the western shoulder of South America, the Andes provide abundant middle-elevation habitat ranging from north of the equator to south of the Tropic of Capricorn. Soaring above lowland jungles, the rugged hillsides are cloaked by cloud forests that represent an ideal home for orchids. Moist air that accumulated water vapor during a 2000-mile journey across the Amazon is pushed by the eastern trade winds over steep Andean slopes. The sudden change in altitude induces clouds to form and wrings precipitation from them, soaking forests below in a gentle drizzle, the perfect watering system for orchids.

The Andes support an immense diversity of orchids for reasons beyond rainfall. These precipitous and deeply dissected mountains offer the archetypal conditions for speciation. Towering ridges serve as geographic barriers to movement of pollinators and seeds. Populations of orchids that become established in deep valleys are separated from neighboring valleys by these ridges, while orchids adapted for higher elevations are isolated from their neighbors by the selfsame valleys. Thus, a topographically complex mountainous forest provides thousands of small micro-habitats in which orchids grow exceptionally well but enjoy little communication with populations living only a few miles away. Over time, they evolve into separate, distinct species, adapting to the subtly different conditions of each location. Further enhancing the mountains' powers of speciation, climate changes rapidly as one ascends even within a single valley and can be sufficient to prevent low-elevation orchids from establishing in the higher and cooler reaches.

The climatic differences and topographic barriers also affect insects, such as the bees, wasps, and flies that often are key pollinators for orchid species. As insects evolve in isolation from their neighbors, they also diverge and become separate species. The proliferation of distinct species of pollinators in mountainous regions accelerates the rate of speciation of orchids, which often rely on a single type of insect to carry pollen among plants. Thanks to these evolutionarily tight relationships, the division of a single fly species into two is also likely to split the orchid it pollinates into two new species, each adapted to attract one of the newly differentiated flies.

The result is a patchwork of tiny range maps, each occupied by a different orchid species. Often these areas are inhabited by plants of a species that occur nowhere else. In scientific parlance, such species are called *endemic*, meaning they are found in only a single location. Isolated from their neighbors, genetically as well as geographically, orchid populations experience *genetic drift*, the random accumulation of inherited modifications that cause the species to diverge from close relatives. As long as pollinators do not carry DNA across the mountains, the species continues to drift and becomes more unique.

In contrast, when orchid populations remain in genetic communication because of the actions of able-bodied pollinators or the absence of geographic barriers, plants continue to trade genes and will not differentiate. However, an intriguing middle ground also exists. In some regions evolution appears to be balanced on the cusp of creating new species. Populations of orchids have begun to diverge from one another, through genetic drift and local adaptation, establishing proto-species. But low levels of cross-pollination continue, allowing a trickle of genes to be shared. In these circumstances, numerous intermediate hybrid forms arise, with flowers and vegetation exhibiting characteristics of the population to which each parent belongs. Populations never quite congeal into firm species but rather participate in swirling gene pools called *hybrid swarms*. Populations within the swarm are mostly distinct but not quite unique enough to be recognized as separate species.

Hybrid swarms are common among epiphytic orchids in the Andes. Their existence is evidence of a dynamic state of evolution in which populations are rapidly adapting to local conditions and exploiting new habitats. High-speed evolution seems to be characteristic of Andean orchids. Geological evidence suggests that the uplift of these mountains is a relatively recent phenomenon, transpiring over just the last 10 million years. In terms of evolution, that is a brief period of time. Thus, the orchids of South America began to diversify quite recently, and under rapidly changing conditions to which hybrid swarms can adapt swiftly. The picture that emerges is of an enormously diverse set of genes shared between orchids occupying similar but fluid environments, which produce shifting and interwoven clans from which true species eventually emerge. The interplay between rapid adaptation and genetic isolation, the former generating new varieties, the latter cementing them as species, is at the heart of what has been termed the Andean engine of biodiversity.

Over the last few million years, this engine has yielded an astonishing number of species, including perhaps one-third of all orchids on planet Earth. The engine of the Andes also is responsible for producing hundreds of frog and hummingbird species: both groups have achieved spectacular diversity in these craggy reaches. Many of these mountain-dwelling species occupy very small areas, a strongly endemic pattern of distribution that offers opportunities as well as challenges to conservationists.

Pessimistically speaking, endemic orchids are at high risk because deforestation of just a few acres can eliminate a species' entire range, driving it instantly to extinction. Countless orchids already have vanished in this way, as widespread forest loss continues to change the face of the Andes. The familiar drivers of deforestation—population growth, insecure land tenure, weak park protection—all share a quotient of the blame. They have been joined by modern threats to Andean habitats, such as mountaintop mining and large-scale hydroelectric projects. Each, while making important contributions to national economies, put endemic populations of orchids at risk.

In contrast, high levels of endemism offer unique conservation benefits. Protection of small areas, even just a few dozen acres, can safeguard the entire population of some endemic species. The same cannot be said of the Amazon rainforest, where millions of square miles must be conserved to protect jaguars, monkeys, and eagles. This revelation underlines the importance of small forest reserves in the Andes, which can contribute significantly to orchid conservation. Indeed, protection of a few hundred acres of cloud forest will save far more unique species than a few thousand acres of rainforest. Conservation organizations have seized on this concept and are actively promoting establishment of an array of parks, reserves, and private protected forests across the exceptionally diverse slopes of the Andes.

For similar reasons, exploration of the guillotine-sharp ridges and plunging valleys of the Andes is exceptionally likely to lead to discovery of new species. Hiding within those isolated sites, cut off from neighboring orchid populations by the inaccessible topography, hundreds of unknown species await. When botanists gain access to new regions, they inevitably encounter novel orchids, sometimes at astonishing rates. In scientific parlance, such areas have steep species-to-effort curves, meaning that investment in a modest amount of searching effort yields large numbers of new species. Thus conservation programs in the Andes bear twin benefits: protecting known restricted-range species, and providing opportunities to discover new species that add to the global diversity of orchids.

At present there are more species of orchids than any other type of flowering plant in the world. Only the daisy family, the Asters, at some 23,000 species, is a close competitor. Why, then, are orchids so diverse? To answer this question, we must investigate how new species come into being.

The most common method understood by biologists is when a group of individuals becomes geographically isolated from the main population, and over time changes gradually in form until it no longer can be recognized as the same species. Called *allopatric speciation*, this process has occurred repeatedly in the Andes, on oceanic islands, and wherever topographic barriers exist.

Gradual changes can be accelerated if unique evolutionary pressures are imposed on the isolated orchids. For example, if they are pollinated by slightly smaller bees than their remote relatives, narrower flowers may evolve. Over time, these flowers might become small enough to exclude the larger bees that pollinate the original population, and the two groups of orchids would be reproductively isolated: no exchange of genes could transpire, and the two types of orchids would effectively be separate species.

In circumstances that are relatively uncommon, variation in pollinator behavior can produce separate orchid species even within a single, continuous geographic area. For example, some orchids in a bee-pollinated population may develop, quite by accident, pale flowers that attract a few moths by night. Over time, the nocturnally visited blossoms begin to specialize on pollination by moths, perhaps by emitting sweet fragrances at night, while the original population retains features that only bees find appealing. The separation of these two populations into distinct species, known as *sympatric speciation*, unfolds in the absence of any topographical isolation.

The appearance of new species, whether by allopatric or sympatric speciation, is counterbalanced by the extinction of existing species. Without extinction, the rise in planetary diversity would be endless; in fact, it has been relatively stable over geological time, albeit punctuated by several spasms of mass extinction (including the current precipitous loss of species). Thus the total diversity of orchids in any given area is the product of both speciation and extinction. Increases in diversity come about because of low rates of extinction or high rates of speciation, or both. In the Andes, as mentioned, cool and damp environments provide orchids with optimal conditions for survival that lead to large population sizes and low rates of extinction. Add to this the exceptionally high rates of speciation and it is easy to comprehend why Andean orchids are so diverse.

Recent studies have examined the origins of orchid diversity in novel ways. Genetic analysis permits researchers to trace the evolutionary path orchids have taken, assess which groups have contributed most to present-day diversity, and speculate on the crucial adaptations that unleashed explosive phases of speciation. The data suggest that epiphytism is the key. Early orchids were terrestrial plants, but a few pioneers eventually made their way, quite by accident, into the treetops and trunks. There, a suite of adaptations emerged that helped orchids survive, and thrive, in this lofty environment: twining roots clasp branches, pseudobulbs store water and nutrients, small plant size and rapid development enable early flowering, and windborne seeds scatter like dust to encounter unoccupied perches.

These adaptations permitted early epiphytic orchids to establish a foothold in their new environment. Variations on the pioneers' basic form yielded a proliferation of species that began to fill the microhabitats available to epiphytes: shady trunks, sunlit canopies, windy sites, wet sites, hefty branches, and slender twigs. Once the door to an epiphytic lifestyle had been opened, orchid diversity multiplied rapidly as plants specialized in myriad ways to fill these many niches. Today, epiphytes comprise more than 70% of all known orchid species.

A parallel hypothesis for the origins of orchid diversity focuses on interactions between flowers and pollinators. The majority of orchids rely on a single species of pollinator; a few may be visited by two or three species, but less than 5% of orchids are served by a large number of pollinators. Such specificity can lead quickly to coevolution, in which flower and pollinator evolve in response to one another. As a consequence, floral structures and pollinator shapes and behaviors dovetail seamlessly. Bees with extra-long tongues, for example, fertilize only flowers with equally long nectar tubes, and variation in one species favors a matching modification in the other. Once established, these one-to-one reproductive relationships isolate the orchid from nearby plants with whom it no longer shares genes, and from which it rapidly diverges into a distinct species.

The spectacular diversity of orchids that we enjoy today likely is the result of interplay between all these theories. Adoption of the epiphytic lifestyle opened the door to a vast, underexplored habitat. Variation within the forest canopy permitted extreme niche diversification. Topographic complexity boosted speciation rates among isolated populations, and coevolution with diverse tropical pollinators cemented the isolation of distinct species. And the rich, warm, wet, and welcoming forests of the tropics nourished healthy populations of these plants in ideal conditions for growth and reproduction.

Although we can understand some of the ways in which orchids have come to be so phenomenally diverse, no one really comprehends the entire picture. Why did orchids produce so many species, while other types of plants that grow in the same mountainous habitats did not? Why are orchids in some low elevation sites without steep slopes also highly diverse? Are rates of speciation in the Andes accelerating, slowing, or holding steady? Many such questions still are being studied by scientists and amateur orchid enthusiasts around the world. Solutions to these conundrums may be uncovered in university laboratories, in private greenhouses and botanical gardens, or in the misty forests of the tropics. Who knows, perhaps your next trip to a lush tropical mountainside will yield new clues that help unravel the fascinating and beautiful mysteries of diversity.

Conservation

Support for conservation of orchids, plants once torn heedlessly by the millions from their native habitats, has grown exponentially over the past few decades. Many orchid societies now feature active conservation committees that provide financial and logistical assistance to orchid protection programs both local and international. Land protection projects focused on orchid-rich habitat have sprung up around the globe. Tourists increasingly visit tropical countries with the goal of enjoying native orchids in the wild, and go to great lengths to explore unspoiled ecosystems in search of these enthralling plants; the fees they pay sponsor local guides and reserves. Donations of time and money to habitat conservation efforts are rising steeply, providing aid to adopt-an-orchid programs, large-scale campaigns for land purchase, and novel incentive systems promoting protection of private lands. The conversion from harvesting to habitat protection did not happen overnight, however, but rather has evolved through a long series of incremental steps.

Orchid collecting has its origins in Europe, particularly in England. While some orchids occur naturally in these temperate regions, most are terrestrial species lacking showy flowers, and they did not capture the fascination of everyday citizens. But as exploration of the tropics boomed in the late 18th and early 19th centuries, accounts of so-called parasitic air plants bearing huge, stunning flowers began to seep into public awareness. Travelogues published by tropical explorers spoke vividly of the amazing blossoms, their diverse beauty and copious abundance, and soon an interest in these mysterious and exotic plants was piqued.

Shortly, the first tropical orchids arrived in England and were coaxed into flowering in nurseries, where they were proudly displayed by greenhouse owners and jealously sought by a small but burgeoning flock of aficionados. Initially, only the exceptionally wealthy possessed the resources necessary to coddle tropical orchids that required protection in expensive, stove-warmed greenhouses. With the arrival of cold-tolerant orchids such as *Cattleya*, however, this hobby of lords and ladies became more widely accessible. Soon, the craze caught fire, and seemingly everyone in England wanted these entrancing flowers in their own home.

Nurserymen like Henry Frederick Sander and Harry Veitch dispatched scores of men to the mountainsides of tropical countries, from Borneo to Brazil, in search of wild sources of the increasingly popular plants. Insatiable demand promised fabulous wealth to those who could guarantee a steady supply of orchids for their warehouses. Sander's St. Albans greenhouses received, stocked, and sold more than a million plants in one decade alone. Such numbers are particularly mind-boggling when one considers the nearly insurmountable challenge of locating and shipping tropical orchids. Their homelands were little explored, barely formed countries rife with war and theft, awash with disease, populated by hostile indigenous groups, and teeming with poisonous animals. Transport by mule, horse, and hired hand from remote forest to shipping port was fraught with danger: landslides or floods consigned entire shipments to destruction, thieves routinely poached the painstakingly collected harvests, and mercurial tropical climates could sear, freeze, rot, or dehydrate even the most well-packed specimens. What plants did reach the coast were loaded into sailing vessels to suffer a six-week crossing of the Atlantic in hot, overcrowded, and airless holds. The number of plants that perished is incalculable, but enough did survive to fuel the orchid fever that held Europe in its sway.

By the mid-1800s, a full-scale siege of tropical forests was underway. Late arrivers decried the wanton destruction wrought by earlier collectors, some of whom deliberately obliterated whole populations of marketable species so as to thwart the commercial opportunities of those that followed. The end result was the decimation of countless acres of primary forest. Few could serve as witnesses to the widespread deforestation, and fewer still would have given it a second thought at that point in time. The world was young, relatively unoccupied by humanity, and most inhabitants of and visitors to the tropics were bent on converting, not conserving, the natural environment.

Gradually the passion for orchid collecting spread, and the availability of orchids grew. More and more species, literally from around the world, were snapped up in an effort by collectors to exploit unique varieties not present in any other greenhouse. A new orchid with a unique form or fragrance stimulated a buying frenzy among consumers with a keen appetite for novelty. This urge to cultivate new species continues to the present day. Such yearning is an expression of a deep-rooted appreciation of biodiversity, of something never before seen, and it forms the roots of today's orchid conservation movement.

Among the deluge of new orchid varieties brought to Europe, special attention was lavished on those extracted from tropical mountains dominated by cool temperatures. These orchids could be cared for more successfully in the temperate climates of most collectors. Soon a flowering orchid became a popular decoration in households, even outside the circles of wealth and privilege. Many common gardeners took up the art of orchid collection, their desire to enjoy the flowers transposed into a devotion to care for the plants themselves. What might have begun with a single plant soon expanded into a handful of specimens on a windowsill, a set of shelves to accommodate a few more plants, and the inevitable greenhouse built to receive mounting numbers of orchids.

By the mid-1900s, the fascination with orchids had dispersed around the world. A nursery in Hawaii pioneered commercial-scale production of a particularly gorgeous variety of large-flowered *Vanda* orchid, and sent enormous shipments to the continental United States. There, the blossoms were handed out at store openings, parades, and other events in a campaign known as "free orchids for every lady." Soon orchids became a default flower for corsages, and the worldwide popularity of these formerly unknown plants was cemented.

Botanical gardens, already with a long history of introducing visitors to plants from far-flung locations, were quick to recognize and capitalize on the trend. Across Europe, and later the United States, formal gardens constructed elaborate greenhouses that maintained carefully controlled environments for orchid collections. Visitors flocked to these displays, which in turn fueled popular demand for orchids. The gardens successfully brought to life the look and feel of faraway tropical locations, and some enthusiasts began to dream of seeing such landscapes for themselves.

The soaring popular interest in orchids was paralleled by the growth of international tourism. Families who had never before left their hometowns began considering visits to sunny beaches, mountainous landscapes, and increasingly, remote tropical countries brought ever closer by the emergence of affordable air travel and improved tourism infrastructure. Orchid lovers embarked on extended tours of the tropics and returned with photographs and stories of their adventures. Often these tales were punctuated by stories of encountering a flowering orchid in its native habitat, something heretofore experienced by only a very few.

The Andes mountains in particular began to attract the attention of international tourists. Hiram Bingham's rediscovery in 1911 of the spectacular Inca ruins at Machu Picchu sparked worldwide interest in Peru. The stunning, snow-capped volcanoes of Ecuador drew visitors northward, as did the appeal of Colombian cities featuring well-preserved colonial architecture. In the Andes, travelers encountered habitat positively dripping with orchids. Moss-covered trees, cool breezes, constant clouds, and high humidity represented the perfect conditions for orchids, an environment many visitors had labored for years to replicate in greenhouses back home.

Meanwhile, orchid societies were springing up around the world. They began as loose associations of like-minded gardeners, trading information, tools and supplies, and occasionally plants. As the popularity of orchids boomed, these clubs morphed into formal organizations with substantial membership and considerable clout. Botanical societies began to address social causes, such as the Gardener's Royal Benevolent Institution in England, of which famed orchid grower Harry Veitch was an early chairman.

These early actions of conscience foreshadowed the support by modern societies for orchid conservation. Today such groups routinely sponsor tours to visit orchid sites, both local and remote. Speakers dazzle members with photographs of rare varieties in their native habitat and hold them spellbound with tales, peppered by references to oxcarts and mudslides, of how those plants were reached. Societies hold fund-raisers and donate substantial support to habitat protection programs. Their members are informed about and involved in orchid conservation projects and enthusiastically contribute their time and energy to protecting these plants and the habitats in which they are found.

The maturation of the orchid conservation movement parallels the trajectory traced by ornithological societies. Bird-watching used to mean shooting fowl in flight and mounting them over a fireplace where they could properly be appreciated, perhaps with a glass of sherry and a cigar. Excursions into the wild were merely preludes to squeezing the trigger. But an appreciation of birds as an integral part of their environment soon grew. John James Audubon's famous imagery depicted birds interacting with the world around them: hunting for fish, eluding predators, or scrabbling for nesting material. Bird enthusiasts began to seek firsthand glimpses of these behaviors. Gradually, the urge to collect birds was supplanted by the desire to observe them, in all their living splendor.

Similarly, orchid aficionados now clamor for real-world encounters with their beloved plants. For some, that desire is simply an opportunity to observe firsthand the exact climatic conditions in which certain species thrive. Others are entranced by the painterly beauty of a forest bedecked with wild orchids in flower. Patient tourists may sit quietly for hours in hopes of observing a rare pollination event or of identifying a new pollinator. Enthusiastic explorers of remote sites may even happen upon a new species, still being discovered with surprising regularity. These devotees continue, of course, to revel in tending and enjoying their home collections; but the drive to see an orchid as a part of the world, rather than a solitary plant isolated from its natural habitat, is palpable.

Ecotourism is no longer a marginal phenomenon, geared only to hardy travelers loaded with vaccinations, machetes, and insect repellent. A multi-billion dollar industry, ecotourism extends to every part of the globe and represents an important source of revenue for nations and small landowners alike. Throughout the tropics, families who own a small patch of forest, a shady canyon, an array of hummingbird feeders, or a photogenic waterfall are trying to support themselves by attracting tourists. In some countries ecotourism constitutes the single largest industry, gainfully employing thousands of people as guides, drivers, and business owners.

Bird-watching tourism already is an enormous business, with innumerable lodges around the world serving steady streams of binocular-wielding birders anxious to be amazed by new species and new behaviors. Scuba diving followed a similar arc, from obscure sport to immensely popular hobby that has supported the creation of, among others, the Great Barrier Reef reserve. Orchid tourism, while lagging behind these more high-profile activities, is experiencing tremendous growth. Particularly within upper elevation forests, in the Andes and elsewhere, hotels and private reserves are emerging that cater to orchidists. Tour packages designed to maximize the number of flowering orchids seen by participants are increasingly popular. Income from visitors finances park guards, fences, signage, and other controls that help safeguard wild orchids. In this way, expanding interest in experiencing orchid habitat firsthand is helping to support its protection.

Other approaches to conservation of orchids and their habitat are yielding significant achievements. Steep, forested slopes, once demeaned as agriculturally useless land, now are being declared national parks by governments of tropical nations, thanks in part to pressure and support from orchid groups. Restrictions on the manipulation, extraction, and sale of orchids, formerly ignored or completely absent, are now being enforced by local authorities. At the international level, the Convention on International Trade in Endangered Species (CITES) has established strict guidelines for the movement of orchids across international borders. Although CITES is justifiably criticized by many within the orchid world, it must be noted that its protections were extended to orchids because of the rapacious nature of collectors, along with the effects of widespread deforestation. Modifications of the Convention are surely required, but the unrestrained free-for-all of the past can no longer be permitted.

As popular support for protection of orchid habitat has deepened, a greater proportion of commercial greenhouses have dedicated themselves to sustainable production methods. Today, nearly all professional orchid nurseries trade exclusively in plants that have been raised from seed, or vegetatively propagated, rather than harvested from nature. Rearing tiny plantlets in glass bottles lifts pressure from wild populations that only a few decades ago were being raided to provide most of the orchids sold in the world. Orchid collectors increasingly take care to verify the sources of their plants, and several high-profile criminal cases have spotlighted the consequences of ignoring national and international law.

As sustainable orchid cultivation practices become mainstream, some may question the validity of protecting habitat. If we can grow under glass all the orchids we need to support businesses and to satisfy enthusiasts, why should resources be spent on conserving tropical forests? The same argument can be extended to botanical gardens, greenhouses, and seed banks. Why protect habitat, if we already have the plants?

The answer arises more from moral conviction than logical proof. Simply observing an isolated orchid with one's eyes, noting its fragrance with one's nose, is a shallow way to appreciate a living organism. An orchid is much more than merely its shape, color, and scent. It is a member of a complex ecosystem, a web of living organisms that interact and inter-depend. The orchid is pollinated by insects or hummingbirds and provides them with nectar. It relies upon, and is a host for, root fungi that are critical to its germination and survival. It clings to tree branches, and plummets to earth if they snap. Mountain breezes waft its seeds to new locations, while clouds exhaled by trees themselves provide moisture. The ecosystem in which it lives provides nutrients, the final remnants of other organisms that lived and died before it. These and infinite other interactions, many yet awaiting discovery, knit the orchid into an enveloping, embracing environment. The diverse linkages are an essential part of the true, complete definition of an orchid. Without its habitat, that orchid is merely a detached unit, an orphan, lacking connection to the greater world, serving only to entertain rather than enrich us.

Thus the full enjoyment of orchids ought to be reflected in more than just the acquisition and display of plants, but in a desire to protect their habitat and the complex, interdependent lives they lead. Visiting such sites, if you are fortunate to do so, can be a life-changing experience. It will broaden your view of orchids and how they survive and flourish in the natural world. It will introduce you to myriad other species, of plants, of birds, of frogs. Even if you never have the opportunity to travel to a distant tropical mountainside, you can support in countless ways the efforts to protect such places. Your assistance will sustain not only orchid populations but also those local people who yearn for a means to prosper without the compulsion brought on by poverty to chop, char, or vend the forest they love.

Through this book, the authors seek to raise awareness and enjoyment of orchids, for experts with years of experience and newcomers alike. *Orchids of Tropical America* was written to provide a deeper understanding of the lives of orchids and the ecological network in which they are woven. We hope that readers will be encouraged to visit some of the alluring ecosystems where these bewitching plants thrive. And we urge everyone, whether you are a collector, a commercial grower, or a botanical novice, to support the conservation of orchids and their native habitat. The orchids, their pollinators, and future generations of orchid lovers will be grateful.

Orchid Identification

Nearly all orchids can be recognized by visible, external characteristics. Although modern taxonomy relies heavily on genetic analysis, most orchids can be distinguished by the shape, arrangement, and texture of their various parts. The broad diversity of forms exhibited within the family gives us the keys to their identification. Precise shape of the lip is particularly helpful, and whether its edges are smooth, lobed, or frilly. Flower orientation also is informative, with blossoms of a minority of species presenting their lip uppermost, above the column, rather than lowermost. Whether stems are swollen into pseudobulbs, or sprout directly from roots, represents a valuable clue. So does the arrangement of leaves, which varies from simple blade-and-stem fronds to overlapping leaves that obscure the stem to tightly splayed clusters that resemble a Japanese fan. Learning to recognize the variation in key characteristics of plants and their flowers will allow you to identify most orchids.

Unlike many flowering plants with which you may be familiar, color is not particularly useful in orchid identification. The hue of a flower may be moderately helpful, but the overall form is far more significant. In the following sections we will outline the range of variation in orchid flower shapes and establish a straightforward language for describing them. Flowers are by far the most revealing part of an orchid, and without them specimens may be difficult or impossible to identify. Evolution and natural selection work first and foremost through reproduction, the manner in which organisms perpetuate their lineage; therefore, reproductive structures are more informative than vegetative parts such as leaves, stems, and roots. Nevertheless, a few broad classes of orchids can be recognized vegetatively, and we will introduce terminology and descriptions for those structures that will be helpful in the early stages of the identification process.

What Is an Orchid?

Orchids belong to a single family of plants, the Orchidaceae. They are monocots, like grasses, palm trees, and lilies, and share many features with these relatives. In their diversity they are truly spellbinding, and among their species one encounters a nearly infinite variety of forms. Nonetheless, the family coheres around several fundamental characteristics. Orchid leaves, usually narrow and strap-like, always have parallel veins running from base to tip. Their flower parts come in threes: an outer whorl of three sepals, plus an inner whorl of three petals, with one of the latter distinctly modified (flared, pouched, or lobed) and referred to as the lip. Distinctive to orchids is the fusion of reproductive parts, the male stamens and female pistils, into a single organ called the *column*. When fertilized, they produce fruits called *capsules* that open to release vast numbers of dust-sized seeds. Virtually all orchids on Earth share these basic attributes.

Move beyond the fundamentals and you are presented with a cornucopia of flower shapes, colors, sizes, and orientations, as well as a plethora of growth forms, from subterranean plants with no leaves to miniatures smaller than a thimble to bamboo-like stems towering 15 feet (4.6 m) tall. In this diversity one discovers the delight experienced by all orchid lovers: every plant presents something unique, something different, something unexpected. And each of these variations, minor or major, has evolved for some specific reason: to attract a novel pollinator, bear the brunt of harsh climates, or deter herbivore pests. In some cases, thanks to careful study, we now understand these ecological connections; in too many species to count, however, we are left with only guesses and the appeal of a mystery yet unsolved.

Like all living organisms, orchids can be classified hierarchically. As plants, they belong to the kingdom Plantae, within which they are one of many families. Individual species of orchids are gathered with close relatives into a group called a *genus* (plural *genera*). A single genus may contain as few as one species (e.g., *Mexipedium*), or more than a thousand (e.g., *Epidendrum*). Regardless of its size, a genus gathers species that are more similar to one another than to orchids in other genera. In standard scientific language, the genus name is capitalized, the species name is lowercase (e.g., *Mexipedium xerophyticum*), and both are italicized. The genus may be abbreviated, so long as the meaning remains clear (e.g., many species of *Epidendrum* are fragrant, such as *E. nocturnum* and *E. falcatum*).

In this book we concentrate on describing the characteristics of the most widespread and commonly encountered genera in tropical America. Coverage of every single species is beyond the scope and aim of this book. Numerous excellent books designed for advanced users, many of them specific to a single country, are available to help identify individual orchids to species. Our purpose here is to provide an introduction to the fundamental differences between genera. We attempt, when possible, to explain why these differences exist by highlighting the evolutionary pressures on reproduction, growth, and survival that produced the adaptations we see today. The goal is to equip readers with broadly useful tools to better understand orchids, without plunging too deeply into the minutiae that separate individual species.

Most people see orchids as an overwhelmingly diverse collection of plants, baffling in their variety and frankly somewhat impenetrable. Upon seeing an orchid, we may recognize it as such but lack deeper knowledge about it: Why is it shaped thus, what makes it different from other orchids, where does it grow, what pollinates its flowers? It is as if we were restricted to recognizing winged, feathered creatures as simply "birds," with no appreciation that hummingbirds hover to feed on flower nectar, that eagles soar to spy prey from above, or that ducks dive to forage on aquatic plants. But if you can recognize a woodpecker, and understand why it pounds its bill against tree trunks, then you possess a deeper understanding of why the bird looks and acts the way it does, even if you cannot identify its exact species.

We seek to provide readers with the same level of awareness of orchids that backyard naturalists have of birds. We not only describe what each genus looks like and how it differs from others but also explain why those shapes exist and what adaptations they represent. After using the book in the field, or reading through it at home, you will acquire a greater understanding not only of the differences between orchid genera but also a deeper appreciation of the ecology and adaptations that set them apart.

Illustrated Glossary of Key Orchid Characteristics

Orchids are most readily and reliably distinguished by their flowers. Only a very few genera can be identified easily with non-floral characteristics alone. For this reason, the genus accounts in this book begin with a description of the flower and its component parts.

An outer trio of sepals and an inner trio of petals make up the most conspicuous elements of an orchid flower (Figure 1). They can be quite plain or highly and distinctively modified. The dorsal sepal is uppermost, held vertically or occasionally *hooded*: cupped with the concave side facing forward. The remaining pair are referred to as lateral sepals. Sepals can be broad or narrow, blunt or pointed, with the tips knobbed like antennae (e.g., *Restrepia*) or highly elongated into trailing threads (e.g., *Dracula*). The sepals can be spread open in a flat plane or curled forward into a cup, bell, or tube.

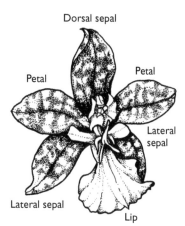

Figure 1. Basic flower form showing sepals, petals, and lip

Occasionally sepals are fused to one another, wholly or partially (Figure 2). The result is a flower that may appear 4-parted if the lateral sepals are entirely fused and the petals are large (e.g., most *Pleurothallis*), 3-parted if sepals are incompletely fused and petals are diminutive (e.g., *Masdevallia, Trisetella*), or even 2-parted if the petals are minuscule and sepals are wholly fused (e.g., a few *Pleurothallis*).

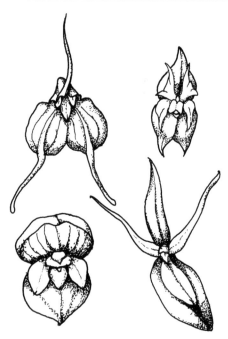

Figure 2. Flowers showing fused sepals

Figure 3. Nectar spur, extending to lower left

The bases of the sepals or petals may be extended into a slender horn projecting behind the flower, known as a *nectar spur* (Figure 3). The spur can be fused with the stem, in which case it goes unnoticed, or it may be free and conspicuous. Long-tongued insects such as butterflies, moths, and some bees, as well as hummingbirds, probe the spur to feed on the sugary reward within.

Petals reside inside the whorl of the sepals; that is, if a flower is held upright, with the stem down, the petals lie atop the sepals. Two of the petals are oriented laterally and may resemble the sepals in shape, color, and size. Those petals can exhibit any of the variations described above: long, thread-like, trailing petals are characteristic, for example, of many *Phragmipedium* (see Figure 6). The third petal, however, is quite different, modified by evolution to play a direct role in the elegant, complex interaction between flower and pollinator that results in fertilization. This third petal is referred to as the lip (see Figure 1).

Variously serving as landing platform, billboard, bucket, trapdoor, or catapult, an orchid lip is perhaps the most intricately modified floral element. Its exact shape, size, texture, and orientation are often the most practical characteristics for identification of orchids. In the simplest case, although quite rare, the lip may be little different from the sepals or petals (e.g., *Platystele*). Or it may be so tiny as to nearly pass unnoticed (e.g., *Stelis*). More frequently, the lip provides a landing platform for visiting insects: flared and outstretched or curled into a tunnel, and appealingly decorated with stripes, spots, and ruffles. It can be deeply notched (Figure 4) or extravagantly fringed (Figure 5), and its texture varies from delicate and transparent to fleshy or waxy to densely hairy.

Commonly, the lip comprises three distinct lobes, one frontal and two lateral, at its base. The relative size of the three lobes, their shapes, and the breadth of the connection between them can be helpfully distinctive. Lateral lobes may be outstretched horizontally, swept back, or held upright like a pair of parentheses between which insect visitors are guided to the reproductive structures. The front lobe may be curled down at the tip, bent at the midpoint, or its sides may be swept upward, forming a concave slide. In some orchids the lip is shallowly pouched, in others slipper-like, and in a few it is deeply concave like a bucket (Figure 6; e.g., *Coryanthes*).

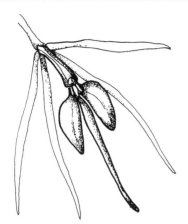

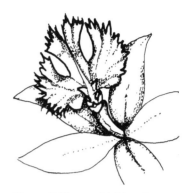

Figure 4. Flower with deeply notched, 3-lobed lip

Figure 5. Flower with fringed (frilly), 3-lobed lip

Reposed on the upper lip surface of most orchids is a raised, fleshy hump, known as a *callus*. This hump, located between the lateral lip lobes if they are present, produces glandular secretions to attract pollinators in many species. Mechanically, it can force insects against the overhanging column as they climb over the hump, or steer the bill of hummingbirds toward the flower's nectar reservoir. The callus exhibits a telling range of forms (Figures 7 and 8). Some are low and broad, others are raised like a craggy mountain, a few are curved into a yoke-like shape. They may be smooth or imprinted with complex textures: parallel ridges, knife-like vanes, jagged teeth, or warty knobs. The exact placement, shape, and texture of this hump can be very useful for identification of genera.

Early in orchid evolution, the lip was the topmost part of the flower; it developed from the upper petal. Once pollination by insects began to dominate, there

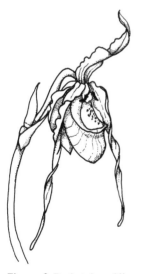

Figure 6. Bucket-shaped lip

Figure 7. Lip detail: shallow, broad side lobes; notched front lobe; keeled callus

Figure 8. Lip detail: deep, wing-like side lobes; ruffled front lobe; warty callus

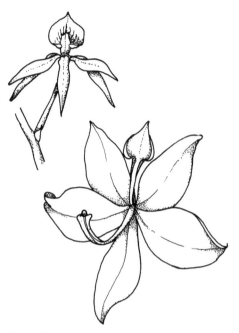

Figure 9. Non-resupinate (lip uppermost) flowers

was strong pressure for the lip to be oriented lowermost, where it could serve as a landing pad for visitors. Most orchids have this orientation, achieved by a process called *resupination*. During flower development, the entire bud twists on its stem 180 degrees, and the emerging flower now has the lip pointed down. A small number of genera, however, have reverted to a lip-uppermost orientation (Figure 9), known as *non-resupinate* (see also Figures 13 and 14). They achieve this either by failing to twist the developing bud, or in a very few, by rotating it an additional 180 degrees. In both cases, the end result is the same—the lip is held above the column. Note that in orchids with hanging inflorescences, or those with drooping flowers, it can be difficult to distinguish resupinate from non-resupinate; examining multiple blossoms is recommended.

As mentioned earlier, reproductive organs of orchid flowers are gathered together in the column, a post or club-like structure (Figure 10). It may be long and slender, short and stubby, or tiny and barely visible. In many flowers, the column is straight, but a distinctive arching form characterizes certain genera. The sides of the column may bear fin-like wings, their shape ranging from narrow keels to broad, curling flaps. Occasionally they are referred to as ears (e.g., *Encyclia*).

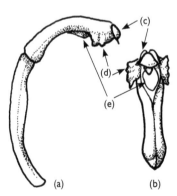

Figure 10. Column detail: (a) side view, (b) front, (c) anther cap, (d) column wings, (e) stigmatic zone

Fused into the column tip are the male structures, known as *anthers* in other plants. While non-orchids produce powdery pollen, in orchid flowers these dust-like particles are coalesced into small beads, often rice-shaped, called *pollinia* (Figure 11). The pollinia are covered by a small hood, the anther cap. The cap shape and placement on the column can help distinguish certain genera (e.g., *Macroclinium*, in which the anther cap is found on the column's dorsal surface). Farther down the column, on its underside, is the female portion of the flower (see Figure 10). Referred to as the stigmatic zone, the stigmatic area, or the receptive zone, this shallow groove or notch-shaped depression receives the pollinia, which are wedged into or smeared on

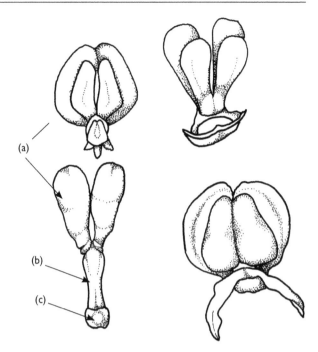

Figure 11. Pollinarium detail: (a) pollinia, (b) stipe, (c) viscid disc

the stigmatic zone by pollinators (or rarely by rainfall), whereupon the packet dissolves and the constituent pollen grains fertilize the flower.

A single pollinium is crammed with thousands and thousands of pollen grains, all delivered in one packet by pollinators. A pollinium typically is mounted on a stalk, or stipe, and attached to a sticky disc called the *viscidium*. Pollinators brush against the disk, and the entire structure, known as a *pollinarium*, is glued to the insect or bird. In this book we refer to the pollen packets only as pollinia, avoiding the technically accurate but more elaborate term pollinarium. Pollinia are arranged in pairs, quartets, or groups of eight, and the exact number is taxonomically informative; however, this examination cannot be conducted without removal of the pollinia, which renders the flower sterile and is strongly discouraged for readers of this book.

Below the stigmatic zone lies the column foot, at the very base of the trunk. This foot occasionally serves as the attachment point for hinged lips that may be delicately balanced (e.g., *Anguloa*) or spring-loaded (e.g., *Porroglossum*).

Beyond these basic flower forms, some truly fantastic variations exist. Sepals, petals, lip, and column may be so elaborately, even grotesquely, modified as to be virtually unrecognizable. Often the sepals and petals are narrow and swept back (Figure 12),

Figure 12. Complex flower (*Gongora*)

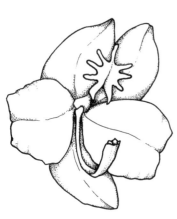

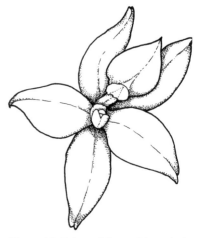

Figure 13. Unisexual flower (*Cycnoches*), male; lip is uppermost

Figure 14. Unisexual flower (*Cycnoches*), female; lip is uppermost

the flower dominated either by a long, arching column (e.g., *Gongora* and *Stanhopea*) or a complex and convoluted lip (e.g., *Coryanthes* and *Mormodes*).

In a few genera, flowers are unisexual: males bear pollen but have no stigmatic zone, while females lack pollinia (Figures 13 and 14). Sometimes male and female flowers appear moderately similar (e.g., *Catasetum*); in other cases the genders bear little resemblance to each other (e.g., *Cycnoches*) and were even classified into separate genera at one time.

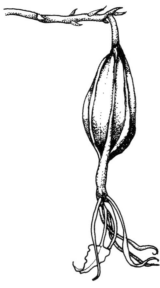

Figure 15. Capsular fruit, showing seams and flower remnants

After fertilization, all orchids, regardless of their flower shape, produce fruits in the form of capsules (Figure 15). A capsule is a specific type of pod that opens, when dry, along more than one seam. Typically these seams are visible as lengthwise ridges on the capsule, which progresses from green to tan to brown as it develops and desiccates. Capsules vary in size and shape, and in number and placement of the seams. Nonetheless, orchids are nettlesome to identify based on fruits alone, so this book describes capsules only infrequently.

A flower is borne by a stalk known as a *pedicel* (Figure 16). If more than one flower arises from a single stalk, even if they are not all open at the same time, the bunch collectively is called an *inflorescence*. In this case, the inflorescence is borne on what now is called a *peduncle*; however, this book avoids these two technical terms, opting instead for the more readily understood word, *stalk*. Inflorescences may be simple or branched. Some

Figure 16. Single flowers, each on a pedicel

Figure 17. Compact inflorescence with flowers tightly packed

Figure 18. Loose inflorescence showing bracts on pedicels

are quite short, while in other genera they twine through the forest for many yards (e.g., *Cyrtochilum*). Individual flowers may be densely clustered (Figure 17) or loosely scattered (Figure 18) along the inflorescence. If the former, they may be arranged on an elongated axis or a more compact, ball-shaped head. Inflorescences may emerge from the base or tip of a pseudobulb (see Figures 25 and 26, respectively), from atop the leaf (see Figure 30), or the center of a whorl of leaves (see Figure 36).

In addition to the flowers, an inflorescence may also bear flaps called *bracts* (Figures 18 and 19). These are modified leaves and often appear green and leaf-like or spike-like. They may be interspersed with the flowers along a loose inflorescence or attached to the base of individual blossoms. Their presence and location can be helpful

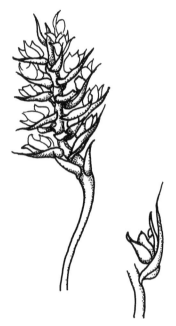

Figure 19. Inflorescence of small flowers with spike-like bracts; detail at lower right

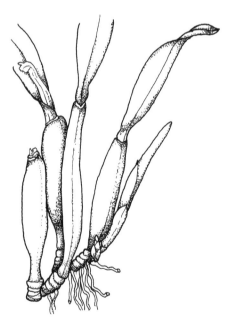

Figure 20. Rhizome (with roots) connecting loosely spaced pseudobulbs, each bearing leaves

in distinguishing similar genera. In other inflorescences they are brilliantly colored and more closely resemble petals. Occasionally these vivid bracts envelop small, pale flowers, dominating the floral display (e.g., *Elleanthus*).

Orchid flowers without a doubt provide the most distinctive characteristics for identifying a particular genus. Broad groups of genera often bear close resemblance vegetatively, however, displaying similar leaves, stems, and pseudobulbs. Thus, an understanding of vegetative, or nonreproductive, characters is essential. In the orchid identification guide below, we concentrate on vegetative attributes to place the plant within a group of similar genera, then rely on floral characteristics to narrow the field to a particular genus.

With few exceptions, all orchids produce stems that bear leaves and roots. Stems often are connected in a series by an underground rhizome, a sort of subterranean stem (Figure 20). As the rhizome is not typically visible, apart from a few cases, this book downplays its importance in identification. Stems may be clustered or loosely spaced along the rhizome. Orchid roots emerging from the stem's base are readily observed in epiphytic species but are usually buried in terrestrials. The color, length, and growth form of roots can help identify genera: some are green and matted (e.g., *Campylocentrum*), others spiky and upturned (e.g., *Gongora, Mormodes*), while still others dangle in the air (e.g., *Huntleya*).

In many genera, the stem is swollen into a bulbous storage organ unique to orchids. Called a *pseudobulb* for its visual similarity to the overwintering bulbs of irises and tulips, its purpose is quite different in orchids. Pseudobulbs store nutrients and water and are particularly common in epiphytic species, which live disconnected from the moist, rich soil far beneath them. The presence of pseudobulbs and their shape are useful aids for classifying an orchid into a small group of similar-looking genera.

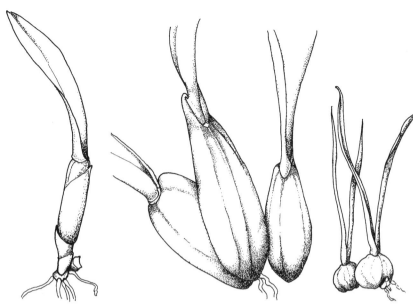

Figure 21. Typical elongate pseudobulb, partly sheathed

Figure 22. Flattened pseudobulbs, from front (left pair) and side (right)

Figure 23. Garlic bulb-shaped pseudobulbs

The simplest pseudobulbs are elongated and vaguely cylindrical. Pseudobulbs may be tightly clustered or widely scattered along the rhizome (see Figure 20). They may be egg-shaped or elongated (Figure 21). In several genera the pseudobulbs are overtly flattened from side to side, as if squeezed in a vise; technically, this form is called *compressed*, but we refer to them as *flattened* (Figure 22). If squashed from top to bottom, a form referred to as *depressed*, the pseudobulb resembles a garlic bulb or a chocolate kiss (Figure 23), and is quite distinctive (e.g., *Encyclia*). Other telling shapes include a cigar (e.g., *Catasetum*), a narrow cylinder (e.g., *Cycnoches*), or hollow with a small hole providing access to a central chamber (e.g., *Myrmecophila*). The color and texture of pseudobulbs also are informative. Some are nearly black (e.g., *Eriopsis*), while others are pale and glossy (e.g., *Batemannia*). Although pseudobulbs typically are smooth sided, some are prominently furrowed (Figure 24), with ridges running vertically (e.g., *Anguloa*).

Figure 24. Furrowed, clustered pseudobulbs, with apical (top) leaves

Leaves may arise from the top, or apex, of a pseudobulb (Figure 25), or they may emerge from both its apex and its base (Figure 26). Basal leaves often clasp the pseudobulb, occasionally

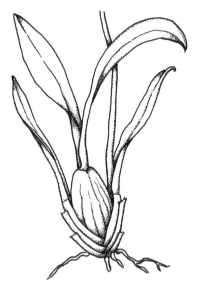

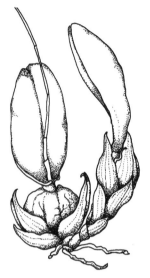

Figure 25. Pseudobulb with apical (top) and basal leaves; inflorescence from base

Figure 26. Pseudobulbs with sheaths; inflorescence from apex (top)

obscuring it. In some genera, apical leaves drop off during the dry season and only their bases remain, sometimes as short spines (e.g., *Cyrtopodium* and its relatives). In other genera, older leaves fall off, leaving distinct seams crossing the pseudobulb; these are known as jointed pseudobulbs (e.g., *Warrea*). Besides leaves, pseudobulbs may be sheathed, partially or entirely, by tissues that vary from leaf-like to fibrous to net-like (e.g., *Isabelia*). Sheaths also may hide the pseudobulb, making it difficult to recognize as such (see Figures 21 and 26). A few genera (e.g., *Bletia, Govenia*) even have subterranean storage organs, called *corms*, from which leaves arise.

In orchids that lack pseudobulbs, variations in the form and texture of the stem are instructive. The stem may be completely covered by overlapping leaf bases (Figure 27; e.g., *Lockhartia, Dichaea*), or visible between them. Stems may be

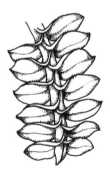

Figure 27. Stem with overlapping, clasping leaves

soft and scrambling, often referred to as prostrate, or they may be tall and reedy, or cane-like (Figure 28; e.g., *Epidendrum*). Stems can be smooth or hairy, occasionally extremely so. Others have very characteristic sheaths, shaped like upturned trumpets (Figure 29; e.g., *Lepanthes, Stelis*), referred to as *lepanthiform* after their characteristic genus.

From the stems or pseudobulbs of nearly all orchids arise leaves, of varied shapes and arrangements. Leaves may be absent during the dry season, or from terrestrial orchids when they are in flower. Like pseudobulbs, leaf characteristics are most helpful for classifying orchids into a relatively small cluster of candidate genera. The first step of the identification system in this book relies heavily on pseudobulb and leaf characters.

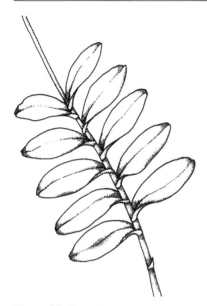

Figure 28. Cane-like stem with leaves alternating, 2-ranked

Figure 29. Stem with trumpet-shaped (lepanthiform) sheaths

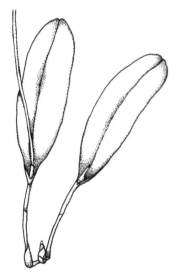

Figure 30. Elliptical leaf shape; inflorescence emerging from leaf base

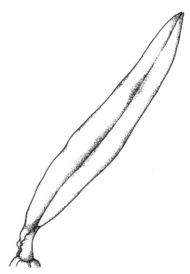

Figure 31. Strap-shaped leaf, with folded base

Orchid leaves typically are elliptical in shape (Figure 30), long and tapering at both ends. The presence at the leaf's base of a distinct stalk, the *petiole*, is a distinctive characteristic of some genera (e.g., *Govenia*). Leaves commonly may be longer, parallel sided and less tapering (Figure 31), a form usually referred to as strap-like (or technically, *linear*). In such leaves, we carefully distinguish

Figure 32. Pleated (plicate) leaf, showing cross-sectional cut through accordion-like texture

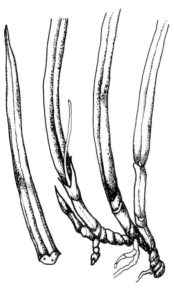

Figure 33. Cylindrical (terete) leaves, showing cross section and central crease

between the terms narrow and thin: "narrow" describes the breadth of the leaf from side to side, its opposites are broad or wide; "thin" characterizes the thickness of the leaf from the upper to lower surface, its opposites are thick or fleshy. Leaves of some orchids are distinctive in having their bases folded into a petiole where they meet the pseudobulb.

Leaves, whether strap-like or broad, exhibit variation in their texture. Leaves that are somewhat spongy, or aloe-like, are called *fleshy*. The term *leathery* applies to stiff leaves. Some are soft textured, a common feature of terrestrial plants. In some leaves the veins, running longitudinally, are quite visible. In many orchids the leaves are distinctly pleated, like the bellows of an accordion (Figure 32). The technical term for this texture is *plicate*. Other leaves have a distinct central fold, or crease, on the upper surface (e.g., *Stanhopea*). Finally, some leaves are so narrow yet thick as to be virtually cylindrical in cross section (Figure 33). The technical term for such leaves is *terete*, and only a few orchid genera have such foliage (e.g., *Brassavola*).

Orchids often have characteristic leaf arrangements, whether a conspicuous stem is absent or present. Visible stems may be short or long and cane-like, the latter shape serving to more widely separate the leaves (e.g., *Epidendrum*). Leaves may arise on only two sides of a stem: in this arrangement, known as *2-ranked*, the leaves are attached on opposite sides from one another, in a single plane (see Figures 27 and 28). If the bases of the orchid's leaves are more condensed, and still 2-ranked, the resulting shape resembles a fan. Fans may be loose and open (Figure 34), or tightly packed (Figure 35); the latter is characteristic of several groups (e.g., *Ornithocephalus*). In terrestrial orchids, leaves commonly are arranged in a whorl or rosette around the stem, but only at its base, where they lie nearly flat on the ground; this configuration is referred to as a *basal whorl* (Figure 36).

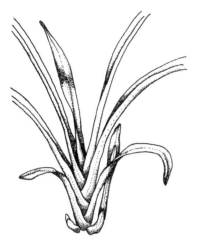

Figure 35. Leaves in tightly packed fan

Figure 34. Leaves in loose fan

Taken together, the characteristics just described are fundamental for the identification of orchids to a single genus. These characters are easily recognized and do not require a magnifying glass or other equipment. With a little practice, you can observe and swiftly recognize the variations in these features. Consider visiting a local florist, a greenhouse, an orchid nursery, or an orchid society show to see and appreciate the wide range of forms that exist within the orchid family. Once comfortable with these characteristics, you will be able to use the identification guide in this book to readily, and accurately, classify most orchids you encounter.

Figure 36. Terrestrial orchid leaves in basal whorl; inflorescence from center

How to Use This Guide

In this book we treat 122 genera of orchids, which include several thousand species. Despite this expansive diversity, however, most plants in these genera should be readily identifiable, by experts and novices alike. Once the basic external characteristics of orchids have been mastered (see the Illustrated Glossary above), accurately placing a plant into one of these genera should be, in nearly all cases, relatively straightforward. Carefully observe all parts of the plant and its flowers, then follow a few short steps in the identification guide below.

The guide begins by organizing the genera into 12 broad groups, each defined by a single characteristic or a small set of characteristics. To start your search, select from the table, below, the group that best matches your plant; for example, *Lip a Deep Pouch*. Next, jump to the page that focuses on that group. There you will find several subgroups distinguished by additional characteristics such as *Leaves in loose fans* or *Stems tall and slender*. Within each subgroup are one or more genera; if more than one, a brief description of how they differ is included. Use the subgroups and descriptions to narrow your orchid to one, or a handful, of candidate genera.

Now turn to the illustrated genus account for each of your candidates. Photographs, while not intended as definitive identification aids, nonetheless provide a quick means of checking a preliminary identification: Does the flower in your hand look similar to, or totally different from, the picture on the page? Remember that as one of the most diverse families on earth, orchids exhibit a huge range of forms. Shapes, sizes, and hues vary widely between genera, and even a single genus can exhibit considerable diversity. Nevertheless, the photographs, when viewed as a group, illustrate the characteristic form of the genus and its most common variations.

Read the written account for a more precise summary of the key features that characterize the genus. In the text accounts, we have tried to express as thoroughly as possible the range of forms found in a genus, particularly when this range is not fully represented by the photographs. Finally, each account includes a brief section providing concise details that help separate the genus from those that are similar. This text should be used in conjunction with the identification guide and will enable you to winnow a few candidates down to a single genus.

Note that for the purposes of this book, we do not attempt to identify orchids to species, a daunting task even for established experts. For species-level identification, a detailed and technical botanical key is required. Such keys are available for many countries in the region covered by this book, and we wholeheartedly recommend them if more detailed identifications are required. (See Bibliography for specific suggestions.)

In summary, observe your plant meticulously, match its characteristics to the identification guide, review the text and photographs for each candidate genus, and use the details regarding similar species to eliminate false contenders. Once you have confidently placed your orchid into a genus, we invite you to read further about its biology and history, deepening your understanding of your orchid's interaction with the world around it.

Identification Guide to Major Orchid Genera

Group	Major Characteristics	Page
1. Lip a Deep Pouch	Flower lip is deeply concave, forming a bucket.	47
2. Pseudobulbs Compressed (Flattened Sideways)	Pseudobulbs are flattened from side to side, not greatly elongated (oval at most).	47
3. Pseudobulbs Cylindrical or Cigar-Shaped	Pseudobulbs are elongated (from football- to pencil-shaped), roughly circular in cross section, often divided by several horizontal seams.	48
4. Pseudobulbs Taller than Wide	Pseudobulbs are elongated, but not cylindrical or cigar-shaped (see 3).	48

Group	Major Characteristics	Page
5. Miscellaneous with Pseudobulbs	Plants with pronounced but average (bulbous) pseudobulbs, flower forms vary (but see 6).	49
6. Leaves Pleated, Mostly with Pseudobulbs	Leaves are distinctly pleated (plicate) in cross section, like accordion bellows.	49
7. Fan of Narrow Leaves	Leaves are arranged in a single plane, like an unfolded paper fan.	50
8. Leaves Overlapping and Alternate	Small leaves, arranged in a single plane along the stem, overlap at their bases; pseudobulbs are absent.	51
9. Flowers with Two or Three Sepals Fused	Lateral sepals, or all three sepals, are partly to wholly fused; pseudobulbs usually absent.	51
10. Pseudobulbs Absent, Typically Terrestrial	Terrestrial plants (and occasional accidental epiphytes) that lack pseudobulbs (but see 11).	52
11. Leaves in a Rosette, or Absent (when Flowering)	Leaves are arranged in a whorl near soil surface; pseudobulbs lacking; typically terrestrial.	52
12. Pseudobulbs Absent, Epiphytic or Terrestrial	Plants of various forms and habits in which pseudobulbs are absent or hidden.	53

1. Lip a Deep Pouch

Flower lip is deeply concave, forming a bucket.

Leaves in loose fans .... *Phragmipedium*
Stems tall and slender .... *Selenipedium*
Epiphytic, inflorescence hanging
    See 3: Pseudobulbs Cylindrical or Cigar-Shaped .... *Coryanthes*
Epiphytic, inflorescence of a single flower
    See 7: Fan of Narrow Leaves .... *Stenia*

2. Pseudobulbs Compressed (Flattened Sideways)

Pseudobulbs are flattened from side to side, not greatly elongated (oval at most).

Lip 3-lobed or triangular, side lobes flaring as wings or arms
    Side lobes arm-like; callus warty; inflorescence not twining .... *Oncidium*
    Side lobes erect, flanking column .... *Promenaea*
    Side lobes wing-like; leaves pleated .... *Rudolfiella*
    Side lobes broad; lip flaring; petals and dorsal sepal antennae-like .... *Psychopsis*
Lip arrowhead- or jawbone-shaped, side lobes not pronounced
    Lip arrowhead-shaped, tip curling down; flowers large; inflorescence long, twining .... *Cyrtochilum*
    Lip arrowhead-shaped with narrow stalk; flowers small, numerous, confetti-like .... *Notylia*
    Lip jawbone-shaped; sepals form inverted T; petals small; forms diverse .... *Maxillaria*
Lip curled, funnel-shaped
    Flowers small; column hooded .... *Cischweinfia*
    Flowers large; column with arms .... *Trichopilia*

Lip flared, side lobes not pronounced

Lip partly fused to column; stem (rhizome) creeping .... *Aspasia*

Lip clawed (stalked), often notched .... *Cuitlauzina*

Lip base with raise linear keels; lower elevations, southeast Brazil ... *Miltonia*

Lip base with small paired horns; Central America and northwest South America .... *Miltoniopsis*

Lateral sepals fused, forming spur .... *Rodriguezia*

Lip very long, lacking side lobes

Lip long, pointy or triangular; sepals elongated, thin; lip not fused to column .... *Brassia*

Two or three sepals partly or wholly fused

Lateral sepals fused; flower resembles little man .... *Gomesa*

All three sepals fused into tube or cone; petal tips reflective blue ... *Trigonidium*

3. Pseudobulbs Cylindrical or Cigar-Shaped

Pseudobulbs are elongated (from football- to pencil-shaped), roughly circular in cross section, often divided by several horizontal seams.

Inflorescence from pseudobulb tip, or near the tip

Leaves cylindrical; flowers large, white .... *Brassavola*

Flower lacks nectar spur; column thin, arched .... *Cycnoches*

Flower with nectar spur .... *Galeandra*

Pseudobulb hollow, with small entryway hole .... *Myrmecophila*

Cane-like stem resembles pseudobulb; see 12: Pseudobulbs Absent, Epiphytic or Terrestrial .... *Barkeria*

Pseudobulbs stacked, chain-like; see 4: Pseudobulbs Taller than Wide .... *Scaphyglottis*

Inflorescence from pseudobulb base

Separate male and female flowers, males with antennae-like projections on column .... *Catasetum*

Lip a deep bucket .... *Coryanthes*

Callus warty; bracts pronounced, colorful .... *Cyrtopodium*

Inflorescence from pseudobulb middle

Lip and column twisted .... *Mormodes*

Pseudobulbs hanging, inflorescence from leaf base .... *Chysis*

4. Pseudobulbs Taller than Wide

Pseudobulbs are elongated, but not cylindrical or cigar-shaped (see 3: Pseudobulbs Cylindrical or Cigar-Shaped).

Sepals narrower than petals

Lip large, showy; few flowers per inflorescence (or many, in some South American species) .... *Cattleya*

Lip less showy; sepals not broad; few flowers on long inflorescence .... *Laelia*

Many flowers per inflorescence; Mexico to Venezuela only .... *Guarianthe*

Sepals and petals narrow

Flowers purple; Brazil .... *Pseudolaelia*

Flowers greenish; Central America .... *Rhyncholaelia*

Flowers fleshy or waxy

Lip partly fused to column .... *Prosthechea*

Mini-bulb visible where leaf meets pseudobulb; flowers orange with

purple margins .... *Eriopsis*

Dwarf plants, flowers tiny

Pseudobulbs stacked, chain-like .... *Scaphyglottis*

Described elsewhere

See 2: Pseudobulbs Compressed .... *Aspasia*

See 3: Pseudobulbs Cylindrical or

Cigar-Shaped .... *Brassavola*

5. Miscellaneous with Pseudobulbs

Plants with pronounced but average (bulbous) pseudobulbs, flower forms vary (but see 6: Leaves Pleated, Mostly with Pseudobulbs).

Flowers with rearward-pointing spur (long or short)

Leaves visibly veined .... *Bifrenaria*

Leaves blade-like; flowers trumpet-shaped .... *Plectrophora*

Flowers tilt downward (nodding) .... *Comparettia*

Flowers large .... *Trichocentrum*

Leaves nearly cylindrical and grooved, flowers large

Lip arrowhead-shaped; see 3: Pseudobulbs Cylindrical or

Cigar-Shaped .... *Brassavola*

Lip edge frilly, upturned .... *Leptotes*

Lip side lobes wrap column; leaves dangling .... *Scuticaria*

Lip with yoke-shaped, ridged callus

Sepals form chin; lip frilly .... *Galeottia*

Sepals not as above .... *Zygopetalum*

Pseudobulbs shaped like garlic head or onion dome

Flowers resemble drooping Dutch hats; inflorescence a tightly packed

spike .... *Polystachya*

Flowers not as above .... *Encyclia*

Miscellaneous

Leaves fleshy; sepals and petals with radiating stripes; flowers

mimic insects .... *Trichoceros*

Leaves narrow; flowers with sac-like spur; pseudobulbs sheathed with

coarse netting .... *Isabelia*

Described elsewhere

See 2: Pseudobulbs Compressed .... *Maxillaria*

6. Leaves Pleated, Mostly with Pseudobulbs

Leaves are distinctly pleated (plicate) in cross section, like accordion bellows.

Stems cane-like
Inflorescence of many flowers; flowers emerge from bracts .... *Elleanthus*
Inflorescence of one to few large flowers .... *Sobralia*
Pseudobulbs present, mostly epiphytic
Lip arrowhead-shaped .... *Paphinia*
Flowers numerous, cupped, sleigh bell-shaped .... *Acineta*
Column elongated, swan-like; one leaf per pseudobulb .... *Polycycnis*
Flowers downward-facing; one leaf per pseudobulb .... *Sievekingia*
Lip complex, with bucket and twin horns .... *Stanhopea*
Lip complex, petals fused to column .... *Gongora*
Lip complex, arrowhead-shaped with twin horns .... *Houlletia*
Pseudobulbs waxy, 4-sided; lateral sepals in-rolled .... *Batemannia*
Sepals and petals similar, fleshy; inflorescence short, hanging ...... *Peristeria*
Pseudobulbs present, mostly terrestrial
Inflorescence of a single flower; lip frilly .... *Sudamerlycaste*
Inflorescence of a single flower; lip not overtly frilly .... *Lycaste*
Flowers few, cupped, goblet-shaped .... *Anguloa*
Inflorescence (of several flowers) hanging or horizontal .... *Lycomormium*
Inflorescence of multiple flowers .... *Xylobium*
Pseudobulbs jointed (divided crosswise into several sections) .... *Warrea*
Pseudobulbs smooth; inflorescence tall, erect .... *Peristeria*
Complex lip, arrowhead-shaped with twin horns (see above) .... *Houlletia*
Pseudobulbs occasionally present, epiphytic or terrestrial
Flowers with violet bars .... *Koellensteinia*
Pseudobulbs replaced by partly buried bulb-like corm .... *Bletia*
Described elsewhere
See 2: Pseudobulbs Compressed .... *Cyrtochilum*
See 2: Pseudobulbs Compressed .... *Rudolfiella*
See 3: Pseudobulbs Cylindrical or Cigar-Shaped .... *Catasetum*
See 3: Pseudobulbs Cylindrical or Cigar-Shaped .... *Chysis*
See 3: Pseudobulbs Cylindrical or Cigar-Shaped .... *Coryanthes*
See 3: Pseudobulbs Cylindrical or Cigar-Shaped .... *Cycnoches*
See 3: Pseudobulbs Cylindrical or Cigar-Shaped .... *Mormodes*
See 11: Leaves in Rosette, or Absent .... *Cranichis*
See 11: Leaves in Rosette, or Absent .... *Govenia*
See 11: Leaves in Rosette, or Absent .... *Ponthieva*

7. Fan of Narrow Leaves

Leaves are arranged in a single plane, like an unfolded paper fan.
Lip flared, flat to concave at midpoint
Column underside with blunt keel; plants large .... *Cochleanthes*
Column underside with pointed tooth; plants small .... *Kefersteinia*
Lip side lobes wrap column .... *Warczewiczella*
Lip trumpet-shaped, sides pinched in; toothed callus on lip
midpoint .... *Chondrorhyncha*
Leaves in loose fan; lip pouched or tubular, sides pinched in .... *Stenia*

Lip flared, convex at midpoint
  Lip with keeled callus, keel edges pointed or bristly .... *Huntleya*
  Lip with keeled callus, keel edges rounded in front .... *Pescatoria*
Lip arrowhead-shaped
  Flowers thin, spidery, lavender to white .... *Macroclinium*
  Flowers small, compact, yellow to white; see 2: Pseudobulbs
  Compressed .... *Notylia*
Flowers small, column tip long and beak-like
  Flowers small, often hairy .... *Ornithocephalus*
Flowers larger than compact plant, lip flat and flared .... *Erycina*

8. *Leaves Overlapping and Alternate*

Small leaves, arranged in a single plane along the stem, overlap at their bases;
pseudobulbs are absent.
  Stem obscured by leaf bases
    Leaves flattened side to side, blade-like .... *Lockhartia*
    Leaves flattened top to bottom, with visible central fold .... *Dichaea*
  Stem visible between leaf bases
    Inflorescence loosely multiple, or single orange-red flower .... *Fernandezia*
    Inflorescence compactly multiple, toothbrush-like .... *Campylocentrum*

9. *Flowers with Two or Three Sepals Fused*

Lateral sepals, or all three sepals, are partly to wholly fused; pseudobulbs usually
absent.
  Flower appears 5-parted, lateral sepals may be partly fused
    Dorsal sepal sometimes fused at base; leaf stems sheathed .... *Myoxanthus*
    Flower resembles a tiny man; see 2: Pseudobulbs
    Compressed .... *Gomesa*
  Flower appears 4-parted (lateral sepals fused)
    Flowers rarely 5-parted; inflorescence of multiple flowers; leaf bases
    broad, stems narrow...*Pleurothallis*
    Sepals and petals usually end in tails; inflorescence of a single flower; leaf
    bases narrow .... *Brachionidium*
    Lateral sepals shoehorn-shaped; petals and dorsal sepal thread-like,
    knobbed .... *Restrepia*
    Inflorescence of a single flower; column tip hooded .... *Barbosella*
    Cascading inflorescence of more than 50 flowers; not all species show
    fusion; see 2: Pseudobulbs Compressed .... *Notylia*
  Flower appears 3-parted (petals small)
    Sepals bent backward near base .... *Dryadella*
    Flowers plate- or bell-shaped; shell-shaped lip resembles tiny mushroom
    cap .... *Dracula*
    Flowers bell-shaped or tubular; tiny lip hidden within flower ..... *Masdevallia*
    Sensitive lip mobile (on contact); petals tiny .... *Porroglossum*
    Stem with trumpet-shaped sheaths; one flower open at a time; petals large,
    2-lobed .... *Lepanthes*

Flowers mostly triangular, petals small; often many flowers open at a time; stem with trumpet-shaped sheaths .... *Stelis*

Petal tips bright blue, red, or purple, resembling eyes; see 2: Pseudobulbs Compressed .... *Trigonidium*

Flower appears 2- or 3-parted (sepals fused and petals small)

Sepals end in three antennae-like tails; lip lowermost .... *Trisetella*

Sepals end in three antennae-like tails; lip uppermost .... *Scaphosepalum*

Miscellaneous

Sepals concave, shoehorn-shaped; flowers densely hairy .... *Dresslerella*

Sepals virtually entirely fused; flower sides with slit-like openings .... *Zootrophion*

Sepals and petals fused; small flowers tubular; leaves grass-like, tufted .... *Isochilus*

Pseudobulbs in chains; inflorescence from pseudobulb base; forms diverse .... *Bulbophyllum*

Described elsewhere

See 12: Pseudobulbs Absent, Epiphytic or Terrestrial..................... *Octomeria*

10. Pseudobulbs Absent, Typically Terrestrial

Terrestrial plants (and occasional accidental epiphytes) that lack pseudobulbs (but see 11).

Flower lip lowermost

Stem a thin cane; lip trumpet-shaped, flaring .... *Epistephium*

Stem soft; lip with paired, mustache-like tips .... *Microchilus*

Leaves waxy, shiny, appearing oiled .... *Liparis*

Flower lip uppermost .... *Malaxis*

Leaves pleated

See 6: Leaves Pleated, Mostly with Pseudobulbs .... *Sobralia*

See 6: Leaves Pleated, Mostly with Pseudobulbs .... *Elleanthus*

11. Leaves in a Rosette, or Absent (when Flowering)

Leaves are arranged in a whorl near soil surface; pseudobulbs lacking; typically terrestrial.

Flower lip uppermost

Lip helmet-shaped; petals not fused to column .... *Cranichis*

Flowers tubular .... *Cyclopogon*

Flowers hairy .... *Gomphichis*

Dorsal sepal forms pseudolip; petals fused to column .... *Ponthieva*

Flower tubular, lip lowermost

Flowers hairy .... *Pelexia*

Flower opening bonnet-shaped .... *Prescottia*

Inflorescence fuzzy; lip curled, fleshy .... *Sarcoglottis*

Nectar spur downward pointing, tip not fused to flower stem .... *Eltroplectris*

Flower spreading
Dorsal sepal forms rowboat-shaped hood over column .... *Habenaria*
Leaves occasionally in pairs, not rosettes; leaves with long stalks
(petioles) .... *Govenia*
Flowers compactly clustered, typically epiphytic
Flowers tubular or spreading .... *Eurystyles*

12. Pseudobulbs Absent, Epiphytic or Terrestrial

Plants of various forms and habits in which pseudobulbs are absent or hidden.
Sepals and petals similar, leaves leathery or fleshy
Lip large, rounded; pseudobulbs obscured .... *Ionopsis*
Flowers miniature .... *Platystele*
Lip ruffled, tubular; plant a climbing vine .... *Vanilla*
Lip tubular; pseudobulbs obscured; see 5: Miscellaneous with
Pseudobulbs .... *Plectrophora*
Flower spurred; pseudobulbs obscured; see 5: Miscellaneous with
Pseudobulbs .... *Trichocentrum*
Sepals and petals similar, leaves on reed-like stem
Flowers emerge in clusters from sheaths atop reeds .... *Octomeria*
Lip fused to column .... *Epidendrum*
Sepals and petals swept back; cane-like stem resembles pseudobulb .... *Barkeria*
Toothbrush-like inflorescence; see 8: Leaves Overlapping and
Alternate .... *Campylocentrum*
Sepals larger than petals
Leaves leathery or fleshy; sepals end in tails; see 9: Flowers with Two or
Three Sepals Fused .... *Scaphosepalum*
Leaves fleshy; stems with trumpet-shaped sheaths; petals tiny,
unlobed .... *Trichosalpinx*
Leaves narrow, grass-like; lip tiny, tongue-shaped .... *Cryptocentrum*
Sepals smaller than petals
Leaves fleshy; stems scrambling; flowers mimic insects .... *Telipogon*

Orchid Genus Accounts

This guide includes 122 of the most common, widespread, and eye-catching genera of orchids in the American tropics. Each genus account provides a text description of the plant and its flowers, relates pertinent aspects of its ecology and history, and presents a selection of photographs to illustrate characteristic features and the range of shapes and colors of its species. Below, we explain the structure of the genus accounts and clarify the terminology used therein.

Structure and Terminology

Each of the following accounts begins with a one-sentence encapsulation of the key distinguishing features of the genus, succeeded by a more detailed description of flower and plant characteristics. Throughout, we have endeavored to use straightforward language and to avoid botanical jargon. The Illustrated Glossary in the previous chapter explains those orchid-specific terms that are essential or unavoidable. The description section concludes with a summary of the chief characteristics that distinguish that genus from similar genera.

A brief note here regarding the pronunciation of orchid genus names is worthwhile. A few simple rules can be broadly applied. The letters "ch" are always pronounced like a hard "k" (e.g., *Chysis* is pronounced "KY-sis"). Like most English words, the accent on scientific names typically falls on the second-to-last syllable; in genus names ending in "ia" (e.g., *Octomeria*), the accent falls on the preceding syllable (i.e., "oc-to-MARE-ee-ah"). A few genera, however, bear particularly tongue-twisting names, and for these the guide specifies an exact pronunciation (e.g., *Warczewiczella* is pronounced "VAR-sheh-vitch-EH-la").

The second paragraph of each account offers information about the distribution and diversity of the genus. This includes mention of their habit, or growth form: epiphytic, terrestrial, or rock dwelling. Orchids that cling to tree trunks or branches are called *epiphytes*; those growing on the ground are referred to as *terrestrial*. Occasionally epiphytic orchids may fall from branches and survive on the forest floor, or terrestrial orchids may become established on accumulations of humus in the canopy, but these exceptions are uncommon. Orchids in a third group grow on exposed rock. The exceptionally challenging habitat provided by these exposed slabs is colonized only by highly adapted, often staggeringly beautiful, orchids. Technically, such orchids are known as *rupicolous*, more commonly as *lithophytes* (literally, rock plants), but are described herein as rock dwellers. In some genus accounts, habit and site preferences are explained in greater detail, such as terrestrial on steep hills, or epiphytic on fine twigs.

The total number of orchid species in the world, and in any given genus, is a constant topic of debate. Taxonomists nicknamed *splitters* contend that different varieties should be established as new species, while those dubbed *lumpers* embrace the opposite view, that considerable variation can exist between individuals that nonetheless belong to the same species. The truth surely lies somewhere between these poles. Our understanding of the relationships between species (and sometimes genera) continues to evolve as new studies clarify what is undeniably a diverse and complex picture. In the genus accounts that follow we strive to present the most recent taxonomic information, but offer only an approximation of the number of species.

Although disputes regarding the number and identity of orchid species in a genus are commonplace, and in a few cases challenge the very existence of a genus, we prefer to refrain from such debates. Our aim is to provide a reference that helps readers distinguish plants reliably in the field or greenhouse, rather than to weigh in on ongoing, often arcane arguments about classification. When necessary, we indicate that splitting of certain genera is being considered, and make an effort to describe each of the relevant subgroups within the genus. In some cases, photo captions may present a putative new genus in parentheses: e.g., *Pleurothallis (Acianthera) ramosa*. Such taxonomic disputes are fortunately rare: fewer than 1 in 20 of the genera we cover. Ultimately, this book is dedicated to recognizing orchids and learning about their ecology, natural history, and often fascinating interactions with humans. For more up-to-date information on the diversity and taxonomic status of a given genus, we refer you to the online orchid resources provided at the end of the book.

Species names are provided for the photographs that follow each genus description. When the exact species in the photo is undetermined, we utilize the standard botanical terms "cf." and "aff." (e.g., *Restrepia* cf. *elegans*). The former, short for the Latin word "confer" (compare with), indicates the species to which the photo appears most similar, based on appearance and geographic location. The latter, an abbreviation of "affinis" (similar to), is employed when the plant in the photograph is most likely a new species, to indicate the name of the most visually similar species.

In numerous cases the geographic range, habitat preferences, or favored altitude of a genus can help to identify its members. It should be recognized, however, that these are necessary generalizations and exceptions will exist. For purposes of this book, low elevation is defined as altitudes below 3300 feet (1000 m); high or upper elevation refers to locations above 8200 feet (2500 m); middle elevation lies between. Many genera contain members spanning a variety of habitat types; in such cases, the genus account indicates the most characteristic environments, utilizing terms like montane (i.e., mountain) forest, wet forest, hill scrub (i.e., dominated by shrubs rather than trees), and seasonally dry forest. The geographic extent of each genus is provided by naming the countries that bracket the range: for example, *Cattleya* is found in Trinidad and from Costa Rica to Brazil and Uruguay. Where appropriate, added details are given, such as the location of the center of diversity of a genus, that is, where most species are found (see map on page 218).

Each account concludes with an ecology and history section. Here you will find natural history stories and scientific studies that help explain the unique appearance of the genus. This section may, for example, elucidate the intricate interactions between flowers and pollinators, describe key adaptations of plants to their habitat, or explain characteristic defense mechanisms against pests. The intent is to present the orchid as a participant in an interwoven ecosystem, and thereby provide a clearer understanding of the distinguishing features—colors, shapes, fragrances—of each genus.

This final section commonly includes historical information pertinent to the genus. The world of orchids has provided a rich and fascinating history of human interactions, and we have tried to mention many of the most significant scientists, botanists, growers, and explorers who have peered, prodded, scaled, and slogged their way through the American tropics. Many of these tales are as colorful and outlandish as the men and women who make up the cast of characters. We hope to imbue the reader, embarking on your own orchid journey, with the same sense of curiosity, joy, and wonder that inspired these early orchid pioneers.

Acineta

Description: Hefty *Acineta* plants favor thick, strong branches from which characteristic waxy, rounded, bell-shaped flowers hang in loose bunches. **Flower:** Spherical blossoms appear only partially open: thick sepals enclose smaller petals in a shape reminiscent of a sleigh bell. Typically yellow or cream-colored, flowers almost always are decorated with wine-colored spots. Side lobes of the fleshy lip sweep upward around the column, while the cupped front portion nearly closes the orb-like chamber. The lip is rigid, to which the genus owes its name (from the Greek word for "non-kinetic" or "immobile"). The thick, waxy column often is hairy near the base. Most species are fragrant. **Plant:** Bulky (3–4" [7.5–10 cm] tall), egg-shaped pseudobulbs are deeply grooved, with 2 to 4 large, pleated leaves emerging from the top. Numerous flowers on a single, unbranched spike (to 20" [50.8 cm] long) dangle beneath the plant. **Similar:** *Anguloa* is terrestrial, with 1 or 2 flowers on an erect inflorescence. *Gongora* plants are much smaller, the flower sporting a highly complex, extended lip. *Peristeria* pseudobulbs are smooth (when young). *Stanhopea* pseudobulbs are squat, inflorescence stalk is narrow (usually pencil thin or smaller), and flowers are not cupped.

Distribution and Diversity: Some 15 species of *Acineta* grow epiphytically in middle to high-altitude wet forests, from southern Mexico to Peru.

Ecology and History: *Acineta* flowers are pollinated by medium- to large-bodied male euglossine bees that crawl into the space beneath the column where they scratch the lip surface to collect fragrance; they utilize the scents in complex mating rituals (see *Paphinia*). After backing out, they trigger the release of pollinia, which are glued to their back. A large hump on the lip forces the bee to scrape its back against the column, like stepping over a suitcase in a stairwell. The bees collect odor substances with their front legs, then back out and hover while transferring the scent to storage areas on the hind legs, then re-enter the flower and start again. A distinct blend of fragrance chemicals is produced by each *Acineta* species: the unique odor attracts only a single type of pollinator, thus ensuring the bee will exclusively visit flowers of that species.

 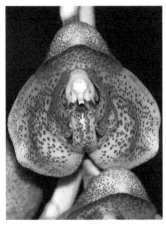 

Acineta antioquiae showing lip side lobes
Andy Phillips

A. superba showing brushy column base
Ron Kaufmann

A. chrysantha showing typical dangling inflorescence
Joe Meisel

Anguloa

Description: Distinctive waxy, goblet-shaped flowers earn *Anguloa* the common name of Tulip Orchid. **Flower:** Broad sepals and petals curve spoon-like around the column, forming a narrow-mouthed cup. Color varies considerably, but white, yellow, and orange are common. A delicately hinged lip is enclosed within the cup: its constant wagging is responsible for another common name, Mother-in-Law's Tongue. The short column bears 2 ear-like flaps near the tip. Long-lived flowers exude strong fragrances by day, including menthol, wintergreen, and cinnamon. **Plant:** Large pseudobulbs (to 7" [17.8 cm] tall) are egg- or football-shaped and usually furrowed, the grooves deepening with age. Soft leaves are long (to 24" [61 cm] or more), broad, and prominently pleated; in high light conditions leaves can be narrow and leathery. Most species, particularly those with yellow flowers, have deciduous leaves; pseudobulbs retain short, defensive spines. Inflorescences bear 1 to 2 flowers, along with several green leaf-like bracts. **Similar:** See *Acineta*. *Sudamerlycaste* and *Lycaste* flowers lack cupped shape.

Distribution and Diversity: The approximately 10 species of *Anguloa* are predominantly terrestrial, although a few are epiphytic. They occur in lower to upper-elevation montane forests, from Venezuela to Bolivia.

Ecology and History: Male euglossine bees visit *Anguloa* flowers, enticed by their odors. Strong fragrance is highly beneficial in tropical forests where dense vegetation forces pollinators to rely on olfactory rather than visual cues. Studies of *A. clowesii* pollination reveal the function of its trap-like flowers. A scent-seeking bee crawls headfirst through the narrow opening, but once inside he discovers the odor exudes from the tip of the lip, behind him. In cramped quarters the bee turns and backs up slightly to gather scent with his forelegs. His weight is now beyond the teeter-totter lip's tipping point, and he must hang on by grasping the petals with his hind legs. When ready to depart he releases this grip, transferring weight to the hinged lip, which springs upward, pressing him against the column where pollinia are now glued to his abdomen. After struggling free, the bee soon is lured to a second flower, where the spring-loaded trap presses him against the column's stigmatic area, completing pollination.

Anguloa clowesii
Ron Kaufmann

A. clowesii
Andy Phillips

A. virginalis showing column wings
Andy Phillips

Aspasia

Description: The pairing of flattened pseudobulbs with flowers sporting a flared, showy lip is characteristic of *Aspasia*. **Flower:** Sepals and petals are narrow, greenish in color with brown barring. Sepals are spread open, while the slightly smaller petals often curl forward. The lengthy column is widest at the tip and is fused to the base of the lip. Brighter in color, the large lip flares broadly but bends down sharply at its midpoint. Flowers usually are fragrant. **Plant:** Pseudobulbs are elongated and flattened side to side, each giving rise to a pair of soft, leathery leaves (to 12" [30 cm] long). The inflorescence arises from the base of a leaf-like sheath above the pseudobulb. **Similar:** *Brassia*, *Miltonia*, and *Oncidium* lack the half-fused, half-folded lip shape.

Distribution and Diversity: Nearly 10 species of *Aspasia* flourish as epiphytes or rock dwellers in low-elevation forest from southern Mexico to Brazil. They prefer the low-light and high-moisture conditions of tree trunks, rather than the canopy or outer branches.

Ecology and History: A false nectary at the base of the lip is one lure for the large euglossine bees that serve as pollinators. Some flowers also emit a pleasant odor to attract *Eulaema* males, who collect the fragrance compounds for use in mating rituals. *Aspasia epidendroides* releases a scent like cinnamon, while *A. principissa* smells of mint or caraway. The latter's principal fragrance component (carvone) is employed by at least 10 other *Eulaema*-pollinated species of orchids, an example of convergent evolution: each orchid found success attracting pollinators with this one odor molecule. Unusually, a case of convergent evolution has occurred among the bees themselves. Black-and-yellow bands mark the abdomens of the *Eulaema* species, much like bumblebees. When the mimics and the model, the bumblebee, are all capable of stinging, their similarity is referred to as Müllerian mimicry. The bright, contrasting colors alert potential predators to the bee's painful sting, a visual warning known as *aposematic coloration*. In Brazil, 3 species (*Eulaema meriana*, *E. bombiformis*, and *E. seabrai*) closely resemble one another, each helping to reinforce the warning message of the shared color pattern.

Aspasia silvana
Andy Phillips

A. epidendroides showing column
base fused to lip
Andy Phillips

A. epidendroides showing flattened
pseudobulb
Ron Kaufmann

Barbosella

Description: *Barbosella* are diminutive plants recognized by a single flower resembling an elongated plus sign. **Flower:** Small (1–3" [2.5–7.5 cm]) blossoms are bland in color, typically green, pale yellow, or purplish, and often translucent. The dorsal sepal is thin and erect, while downward pointing lateral sepals are fused for most or all of their length. Shorter, narrower petals jut out at right angles, or droop slightly. The tiny lip attaches to the column by a ball-and-socket joint, allowing it to swing upward. The stubby, curved column bears narrow wings and a conspicuously hooded tip. **Plant:** Leathery, narrow, grass-like or rarely paddle-shaped leaves are tiny (1/2–2" [1.3–5 cm]); mounted on short stems, leaf bases are wrapped by sheaths that become tan and papery with age. Pseudobulbs are absent. Inflorescences bear 1 flower, but may be displayed in clusters. **Similar:** *Dryadella* sepals bend back near the base. *Pleurothallis* usually bears multiple flowers per inflorescence and lacks very narrow leaves.

Distribution and Diversity: Some 20 epiphytic and terrestrial species of *Barbosella* inhabit cool and wet mountain forests from low to high altitude. These tiny gems can be found in the Caribbean, and from Mexico to Brazil and Argentina.

Ecology and History: *Barbosella* flowers likely are pollinated by tiny flies, the jointed lip trapping them against the column when they land. The plants are so small that some species, like *B. prorepens*, are nearly undetectable in the moss among which they grow. The genus is named for talented Brazilian botanist João Barbosa Rodrigues (1842–1909), past director of the Rio de Janeiro Botanical Garden. His outstretched, pointed mustache and long goatee even call to mind the little flowers of his namesake. An early voyage to study palms in the Amazon (where he described 166 new species) left him desperate to explore the rainforest further. His appeals were answered by his godmother Princess Isabela (see *Isabelia*), who named him inaugural director of a botanical museum in Manaus. His early orchid work was reproduced in the *Flora Brasiliensis* by Alfred Cogniaux, for whom *B. cogniauxiana* is named. During his celebrated career Rodrigues described more than 540 new species of orchids and 28 new genera.

Barbosella dolichorhiza showing fused sepals
Andy Phillips

B. cogniauxiana showing leathery leaves and narrow petals
Ron Parsons

B. prorepens and hooded column
Ron Parsons

Barkeria

Description: *Barkeria* plants are reed-like, with nodding flowers loosely clustered at the top. **Flower:** The purple, lilac, or white flowers have similar sepals and petals that flare back slightly. The broad lip is flat, often pointed, and marked with a large splotch of white or yellow. The column has distinctive fleshy wings, like a stingray. The column is tightly pressed to the lip, but fused only near the base. **Plant:** Stems resemble canes or reeds, are sheathed at the base, and often branch in the middle. Broad leaves with pointed tips emerge from opposing sides of the stem. During dry periods these leaves drop off. Inflorescences arise from the top of a cane and are roughly pyramid-shaped. **Similar:** *Epidendrum* flowers have the column and lip fully fused.

Distribution and Diversity: Approximately 15 species of *Barkeria* occur. They grow as epiphytes, occasionally on rocks, in middle-elevation seasonally dry forest. The genus ranges from Mexico to Panama.

Ecology and History: *Barkeria lindleyana* is pollinated by large-bodied carpenter bees (*Xylocopa*). They alight on the lip, then shove their way between the lip and the column to reach nectar in the flower's base. Only by forcing themselves beneath the column will they pick up and deposit pollinia. Some bees cheat pollination, however, by gnawing their way to the nectar from outside the flower rather than entering it. The naming of *B. lindleyana* brought together three giants in the orchid world. The plant was collected by George Skinner (see *Lycaste*), who sent it to James Bateman (see *Batemannia*), who named it after John Lindley. Lindley (1799–1865) is considered the "father of modern orchidology" for his colossal contributions to the field, including the first sophisticated orchid classification system. A prodigious writer, he published *Genera and Species of Orchidaceous Plants*, which describes 2000 species and required 10 years to complete. He prompted the "Orchids for the Millions" articles that catapulted orchids into mainstream life. As his reputation grew, torrents of plants were sent to him for identification, and in his lifetime he named more than 1940 species.

Barkeria lindleyana
Andy Phillips

B. halbingeri showing column wings
Andy Phillips

Batemannia

Description: *Batemannia* plants combine purplish-brown flowers with white lips, and 4-sided, waxy pseudobulbs. **Flower:** The dorsal sepal is hooded, but otherwise similar to the spreading petals. Downward-pointing lateral sepals are narrower and in-rolled, their bases united forming a prominent chin. Wine-red to brown colors dominate, solidly or patterned on a light base; sepal and petal tips are a pale green. The white lip is spreading and curled down. Its large side lobes, their edges smooth or shallowly jagged, flank an elbow-shaped column. Some species exude a faint, sickly sweet odor. **Plant:** Distinctive pseudobulbs are tall, 4-sided, and shiny. Wrapped by deciduous sheaths, they give rise to 1 or a few leaves. Leaves are thin, somewhat brittle, and pleated, arising from short petioles. The pendant inflorescence, shorter than the leaves and bearing several bracts, supports a loose cluster of flowers. **Similar:** *Galeottia* lateral lip lobes are deeply frilly. *Zygopetalum* sepals do not form a chin; neither have 4-sided pseudobulbs.

Distribution and Diversity: Five epiphytic species favor low- to middle-altitude wet forests. They can be found from Venezuela and Colombia to Bolivia and Brazil.

Ecology and History: Large-bodied euglossine bees pollinate some *Batemannia* species. Presumably the lip acts as a landing pad, its side lobes leading the bee beneath the column to receive pollinia. The genus takes its name from James Bateman (1811–1897), author of "the holy grail of orchid books" (Ferry 2007), the *Orchidaceae of Mexico and Guatemala*. Famed as the largest botanical work ever published (yet only 27" [68.5 cm] tall), it included 40 breathtaking color illustrations. It was so expensive to print that only 125 copies were produced for wealthy and royal subscribers. Now in enormous demand, an original folio recently fetched a quarter-million dollars at auction. Bateman's floral inspirations were delivered by his colleague George Skinner (see *Lycaste*), whose bountiful shipments of Guatemalan orchids supplied most of the featured species. Bateman also was a dedicated grower and a pioneer in cultivation of cool-climate orchids previously consigned to certain death in superheated English greenhouses known as *stoves*. He co-invented a climate-controlled chamber, known as a Wardian Case, still widely used today.

Batemannia colleyi
Ron Kaufmann

B. colleyi showing angled pseudobulb (upper left)
Ron Kaufmann

Bifrenaria

Description: Clustered pseudobulbs, usually with 4 sides, distinctly veined leaves, and spurred flowers characterize *Bifrenaria*. **Flower:** The broad, fleshy flowers (to 2–3" [5–7.5 cm]) are highly variable in color and powerfully perfumed. Lateral sepals are perpendicular to the dorsal and form a rearward pointing spur. Smaller petals sometimes have undulating edges. Lateral lobes of the large, usually hairy lip partially wrap the column, while the front section bends down. **Plant:** Most species have 4-cornered pseudobulbs bearing a single leathery leaf (to 18" [46 cm] long) with overtly raised veins; often a black ring is visible where leaf and pseudo-bulb meet. A few species have 2 thin leaves per pseudobulb. Typically, several flowers are borne on each inflorescence. **Similar:** *Bulbophyllum* leaves lack distinct veins. *Maxillaria* inflorescence bears a single flower. *Rudolfiella* lip has a very narrow base and distinct lateral lobes.

Distribution and Diversity: Some 20 species grow as epiphytes, or rarely terrestrials, in low-elevation wet forest, from Trinidad to Bolivia and Brazil.

Ecology and History: The strongly fragrant flowers of *Bifrenaria* are pollinated by euglossine bees that carry pollinia attached to their heads. The pollinia are mounted on uniquely paired stalks called *stipes* (most orchids have only one), from which the name *Bifrenaria* is derived: Greek for "2 straps." *Bifrenaria steyermarkii* was discovered by Julian Steyermark (1909–1988), credited by the *Guinness Book of World Records* as the champion of plant collectors. He amassed some 138,000 botanical specimens, an astonishing average of 7 per day during his adult life. Numerous species are named for him, including the well-known *Bulbophyllum steyermarkii*. Author of the *Flora of Guatemala* and the *Flora of Venezuela*, he was entranced by Venezuela's tepui mountains. Lofty tabletop mesas rising sharply from the jungle below, tepuis are spatially and genetically isolated from one another. Their unique ecosystems prompted Arthur Conan Doyle to pen *The Lost World*, a 1912 novel about a remote plateau where dinosaurs yet live. A passionate supporter of conservation, Steyermark ceaselessly urged Venezuela to establish nature reserves. His lobbying and significant botanical discoveries led to creation of the Delta del Orinoco National Park, protecting more than 1200 square miles of this spectacular river's jungle waterways.

Bifrenaria aureofulva showing shallow spurs
Andy Phillips

B. tetragona showing pseudobulbs, black rings, and veined leaf
Ron Kaufmann

B. charlesworthii in habitat
Ron Kaufmann

Bletia

Description: *Bletia* plants are easily recognized by the showy flowers and pleated leaves emerging from gladiolus-like bulbs (aka corms). **Flower:** Bright pink or purple colors dominate the similar sepals and petals; a few species are stunning yellow or red. Flowers may be spread open or be partly closed. The 3-lobed lip bears distinctive rows of fleshy teeth or undulating keels, often strikingly colored. Lateral lobes are broad, encircling the curved column; the middle lobe often is curled or wavy. **Plant:** Soft leaves with pleats or visible veins can reach 3 feet (0.9 m) in length. They emerge from corms: underground storage organs that partially protrude from the soil surface. Foliage drops off during dry seasons and may be absent when the plant is flowering, leaving only a long, wiry inflorescence bearing several blossoms.

Distribution and Diversity: Nearly 40 species of *Bletia* live terrestrially on grassy or shrubby, lower- to middle-elevation hills in seasonally dry regions. Their wide range stretches from south Florida to Bolivia and Argentina.

Ecology and History: *Bletia* are pollinated by bees (including *Xylocopa* and *Eulaema* species) that pass beneath the column and pollinia as they crawl deep into the flower. If pollination fails, many species can fertilize themselves. Rainfall may assist by dissolving pollinia and washing pollen grains onto the stigmatic area. Like many deciduous plants, *Bletia* rely on storage organs to survive the dry season. These peculiar corms have been used in myriad ways. Aztecs utilized mucilage from *B. campanulata* to glue feathers to royal clothing, bind paint to fabric, and manufacture corn stalk sculptures. *Bletia purpurea* extract is still used to repair guitars and violins in Mexico and to treat food poisoning in the Caribbean. Flour from pounded corms of either species is the main ingredient in salep, a sort of chewy, room-temperature ice cream eaten with a knife and fork. Salep is best known from Turkey, where it has been concocted from *Orchis* corms for more than 500 years. Its odd flavor has been described unkindly as "a hint of mushrooms, yak butter, or goats on a rainy day" (Hansen 1997: 92). As a warm beverage, salep was widely consumed in England until displaced by coffee in the 17th century.

Bletia reflexa
Andy Phillips

B. purpurea (pale form) showing lip keels
Andy Phillips

Brachionidium

Description: *Brachionidium* means tiny arm, a reference to the small protuberances on the column. **Flower:** Resembling *Pleurothallis* flowers, the delicate blossoms are open only briefly. Sepals and petals are similar, variable in color, and taper to thin tails. Lateral sepals are fully fused. The truncated, waxy lip is wide and hinged to the stubby column. **Plant:** Lacking pseudobulbs these small plants produce shiny, leathery leaves atop short stems. Overlapping, funnel-shaped sheaths with long points punctuate the stems. The single-flowered inflorescence may be nearly erect or may hang downward. **Similar:** *Platystele* lateral sepals are not fused. *Pleurothallis* leaves have long stems, broad bases.

Distribution and Diversity: More than 70 epiphytic and terrestrial species favor middle- to high-altitude wet cloud forests. They can be found from Costa Rica and the Caribbean to Bolivia and Brazil.

Ecology and History: Like many near relatives, the flowers of *Brachionidium* are presumed to be fly-pollinated, the hinged lip likely pinning them against the column. Their preference is for cool to cold forests that would render bees, which are much less active at high altitude, ineffective as agents of fertilization. The exceptionally short-lived flowers, however, present researchers with significant obstacles. First, mature flowers are rarely encountered in the wild, and thus *in situ* studies of pollination are hampered. Furthermore, growing these finicky orchids in greenhouses, where pollination details might be elucidated, is virtually impossible. Even the basic taxonomy of *Brachionidium* species is in question for lack of adequate samples. The formal process of orchid identification or naming follows a meticulous and rigid sequence of steps. Plants collected in the wild are pressed, dried, and adhered to paper sheets for storage in an herbarium. Three-dimensional details of flowers can be studied in specimens suspended in vials of alcohol and glycerin. These samples are reviewed by taxonomic experts and compared with other specimens to determine if the orchid in hand is a new species or a representative of one already described. The delicate consistency of *Brachionidium* flowers, however, impairs their long-term preservation: blossoms wilt away on specimen sheets and can dissolve entirely when stored in vials.

Brachionidium flower, leaves, and funnel-sheathed stems
Ron Kaufmann

B. cf. *ballatrix* showing spade-like column arms
Andy Phillips

Brassavola

Description: *Brassavola* are easily recognized by the large (typically 2–6" [5–15 cm]), long-lived, white flowers and narrow, nearly cylindrical leaves. **Flower:** Slender petals and sepals taper to a point. The large white lip is heart-shaped and pointed at the tip, and in some species highly elongated. The lip's yellowish-green base curls around a short, club-like column that bears small wings. Hidden within each flower stem is a long nectar tube. **Plant:** Slender, stem-like pseudobulbs (to 12" [30 cm] long) are wrapped at the base by whitish, papery sheaths. Each pseudobulb produces a single leathery leaf, very narrow to round in cross section (*terete*); distinguishing pseudobulb from leaf can be difficult. Long leaves (to 24" [61 cm]) typically droop down, but sometimes are erect. Inflorescences bear 1 or a few flowers; rare specimens produce many flowers. **Similar:** *Cattleya*, *Laelia*, and *Ryncholaelia* lack cylindrical leaves.

Distribution and Diversity: Approximately 20 species of *Brassavola* grow epiphytically, or occasionally on rocks (i.e., lithophytically). Some colonize coastal forests and mangroves, others low- to middle-elevation montane forests. They occur from Mexico and Jamaica to Bolivia and Brazil.

Ecology and History: *Brassavola* are pollinated by large sphinx moths. Moth-pollinated flowers converge on a set of characteristics known as a *syndrome*. Large and white, they are readily visible at night. A sweet, musty, or vegetable-like fragrance is emitted, often only nocturnally. Spreading flowers are held away from the plant, accessible to hovering moths; butterflies, in contrast, alight on flowers. Nectar is produced in a deep tube, forcing a moth to press its head against the pollinia while feeding with its tongue, or proboscis. Tube depth and proboscis length often are closely matched. Charles Darwin, upon seeing the incredibly long (10–16" [25–41 cm]) nectar tube of Madagascar's *Angraecum sesquipedale* orchid hypothesized a long-tongued moth as its pollinator. He proposed an evolutionary race between moth and flower: moths develop longer tongues to reach the nectar, whereupon the plant evolves a deeper tube to ensure the moth contacts the column, and so on. Although Darwin's hypothesis initially was ridiculed, a moth (*Xanthopan morganii*) with an enormous proboscis was discovered on the island some 40 years later.

Brassavola perrinii
Andy Phillips

B. nodosa flowers and nearly cylindrical leaves
Andy Phillips

B. reginae
Ron Kaufmann

Brassia

Description: The Spider Orchids of *Brassia* are readily identified by the elongated, thin sepals that hang like arachnid legs. **Flower:** Petals, sepals, and lip are yellow, occasionally lime green, and usually marked with reddish-brown splotches. Sepals (2–12" [5–30 cm]) are longer than petals, the latter often pointed and curling inward. A relatively broad, rounded, or arrowhead-shaped lip bears 2 fleshy ridges on its base that form a narrow groove pointing toward the short column. **Plant:** Egg-shaped or elongated pseudobulbs are usually highly flattened, their base clasped by opposing leafy sheaths. One to 3 leaves, with prominent veins beneath, arise from the pseudobulb tip; an inflorescence of up to 20 flowers arises from the base. **Similar:** See *Aspasia*.

Distribution and Diversity: More than 30 epiphytic and terrestrial species occupy a variety of low- to middle-elevation habitats. They can be observed from the southern United States, the Caribbean, and Mexico to Brazil and Bolivia.

Ecology and History: *Brassia* flowers are unique in that they are pollinated by a group of large wasps called tarantula hawks (*Pepsis* and other pompilids). These sleek spider hunters, with their metallic blue armor and huge size (to 2" [5 cm]), are fearsome insects and their sting is excruciatingly painful. Female tarantula hawks hunt a variety of large spiders, not just tarantulas. The victim is stung into submission, but not death, and hauled to an underground burrow. These powerful wasps are capable of dragging spiders four times their size. Once inside the den the wasp lays a single egg on the spider, then seals the entry hole. A few weeks later a larval wasp hatches and gnaws into the spider's body, which it devours from the inside out. When the larva has fed sufficiently, it metamorphoses into a winged adult and digs its way out. Tarantula hawks apparently mistake *Brassia* flowers for living spiders (an orchid pollination trick known as *pseudoparasitism*), repeatedly stinging the column and attempting to drag the flower away. As it grapples with the lip, the wasp shoves her head against the column, picking up pollinia. Later, at another flower, those pollinia contact the fertile stigmatic area on the column's base, and the flower is pollinated.

Brassia signata showing twin lip ridges
Andy Phillips

B. villosa
Joe Meisel

Brassia verrucosa
Andy Phillips

B. andina, flowers do not open fully
Joe Meisel

B. arcuigera with elongated, flattened pseudobulbs
Ron Kaufmann

Bulbophyllum

Description: The enormous diversity of *Bulbophyllum* defies easy summary; how-ever, most species in tropical America have pseudobulbs arranged in a chain and spike-like inflorescences packed with small flowers. **Flower:** Brownish-purple hues dominate, with blotching and striping common on all parts. Lateral sepals usually are partly fused, but often end in elongated tails. Petals are much smaller but similarly marked. The small lip is partially enclosed by the sepals and deli-cately hinged to the short column, swaying in the slightest breeze. Flowers are borne on a dense stalk, with overlapping bracts resembling a rattlesnake tail. Most species smell of rotting flesh or other unpleasant stenches. **Plant:** Egg-shaped, frequently 4-sided pseudobulbs are evenly spaced, sometimes tightly. One or 2 fleshy leaves arise from the tip, the inflorescence from the base. **Similar:** See *Bifrenaria*. *Polystachya* inflorescence arises from the pseudobulb apex.

Distribution and Diversity: Some 2000 species of *Bulbophyllum* inhabit wet or cloudy forests on all tropical continents, with the center of diversity in New Guinea. Of the few species found in tropical America, most occur only in Brazil.

Ecology and History: Molecular clock evidence reveals *Bulbophyllum* arose on the supercontinent Gondwana before it fractured into Asia, Africa, Australia, and North and South America. The resulting worldwide distribution is therefore explained not by dispersal but by what is called *vicariance*: the land moved, rather than the plants. Most *Bulbophyllum* species are adapted to one of two types of fly pollination. *Myophily* is the attraction of fruit flies and hover flies to nectar and pollen or fragrance. Fragrances may serve as precursors of sexual hormones, or to generate scents that deter predators. *Sapromyophily* involves luring carrion flies to flowers that mimic egg-laying sites like rotting carcasses and decaying vegetation. Such flowers exude rank odors that have been described variously as "all the foul smells imaginable including some new ones" (van der Pijl and Dodson 1966), or more succinctly, like "a herd of dead elephants" (Pridgeon 2006: 42). In both groups the hinged lip traps the pollinator against the column. Some Brazilian species even require a wind gust to trigger the lip, offering flies nectar as a delaying tactic to extend their visit until the wind cooperates.

Bulbophyllum pachyrachis; Florida to Bolivia   *B. cribbianum*
Franco Pupulin   Ron Kaufmann

Campylocentrum

Description: *Campylocentrum* species are characterized by toothbrush-like inflorescences, and either elongated stems covered with 2-ranked leaves or masses of greenish-gray roots. **Flower:** Exceedingly small flowers appear symmetrically 6-parted, as sepals, petals, and lip all are similar. Typically white or yellowish, they are densely packed onto the inflorescence, in 2 files, all facing the same direction. **Plant:** The majority of species bear leaves in a single plane on opposite sides of a reed-like stem (like a fish backbone), a growth form known as *monopodial*. Others are leafless, growing as a tiny stem that sprouts a copious mass of green, tangled roots. When present, leaves are leathery and creased, often with 2 unequal lobes at the tip.

Distribution and Diversity: More than 60 epiphytes and rock dwellers favor forest and cultivated trees, from low to middle elevation. They occur from Florida, the Caribbean, and Mexico to Brazil and Argentina.

Ecology and History: Small bees pollinate many *Campylocentrum* species, clasping the inflorescence and probing the miniature flowers with their mouthparts, to which pollinia adhere. Other species emit nocturnal fragrances that likely attract moths. The genus' monopodial growth form is widespread only in Africa and Asia, and is quite rare in tropical America. Monopodial orchids produce new growth only atop the existing stem (although side branches occasionally develop). In the *sympodial* form, nearly ubiquitous in western hemisphere orchids, new growth is added laterally as subsequent stems spring from a horizontal runner, or rhizome. Orchids with monopodial growth lack pseudobulbs, whereas most sympodial species produce pseudobulbs and rely on them for nutrient storage. Those nutrients are mobilized to the new growth of a sympodial plant, which yields a fresh pseudobulb; the old pseudobulb is called the *back bulb*. Monopodial species, especially those in dry climates, depend instead on fleshy leaves and thick roots for storage. *Campylocentrum* plants rely even further on their extensive roots: the green tissue is photosynthetic and in leafless species provides all the plant's energy. This genus is closely related to African genera that survive in scorching hot, dry conditions; leaflessness in both groups is an adaptation to reduce water loss.

Campylocentrum grisebachii
Ron Kaufmann

C. spannagelii showing monopodial growth
Ron Kaufmann

C. colombianum toothbrush-like inflorescence
Ecuagenera

Catasetum

Description: *Catasetum* derives from the Greek for "downward-pointing bristles," referring to the distinctive antennae-like filaments on the column. **Flower:** All species are fragrant, with dissimilar male and female flowers. Male colors are highly variable but are typically cream or green with maroon markings. The lip is concave, either shallow and frilly, or deep like a wooden clog. Less colorful females have a shorter column lacking the male's antennae. Produced only in full sun, female flowers have a fleshier and more concave lip. **Plant:** Pseudobulbs are cylindrical and tall (to 8" [20 cm] or more), ringed by several horizontal seams, covered by leafy sheaths, and bearing spines. Many broad, pleated leaves emerge from each bulb, dropping off in response to dry seasons. The inflorescence, typically bearing flowers of only one gender, emerges from the pseudobulb base. **Similar:** *Cycnoches* inflorescences emerge near the pseudobulb apex.

Distribution and Diversity: More than 180 mostly epiphytic species prefer low- to middle-altitude seasonally dry forests, from Trinidad and Mexico to Bolivia.

Ecology and History: Charles Darwin called *Catasetum* flowers "the most remarkable of all orchids" (Darwin 1877: 178). He discovered their antennae are triggers for firing pollinia at visiting insects. Male euglossine bees scrabble at the concave lip to collect fragrance. On male flowers they "could hardly gnaw on any part of the great cavity without touching one of the two antennae" (195). Pollinia are mounted on a stalk (stipe), its base a sticky disc (viscidium); the three parts make up a pollinarium ("pollinium" to Darwin). The elastic stipe is bent over a bump in the column, tautly spring-loaded. Triggering the antennae frees the viscidium, and the stipe explosively flings the pollinarium. "I touched the antennae of *C. callosum* whilst holding the flower at about a yard's distance from the window, and the pollinium hit the pane of glass" (186). The structure flies with the viscidium foremost, sticking firmly and violently to the bee's back. Darwin conjectured that, "the insect, disturbed by so sharp a blow . . . flies away to a female plant" (179), and may avoid pugnacious male flowers altogether. When this bee visits a female flower his abdomen dangles beneath the lip, and the pollinia fertilize the stigmatic cavity.

Catasetum tenebrosum male flower
Andy Phillips

C. tenebrosum female flowers
Andy Phillips

Catasetum incurvum male flower
Andy Phillips

C. incurvum female flower
Andy Phillips

C. fimbriatum with hanging inflorescence
Ron Kaufmann

C. expansum showing pseudobulbs
Joe Meisel

Cattleya

Description: Dazzling large flowers with trumpet-shaped lips, thick leaves, and elongated pseudobulbs make *Cattleya* easy to recognize. **Flower:** Large flowers are variously hued, commonly purple, white, or brown. Petals are considerably wider than sepals, often with undulating edges. The large lip curls into a tube at the base, while the flared opening exhibits a notched and wavy margin. **Plant:** Pseudobulbous stems are thick and club-like or columnar, often fluted, with horizontal joints. Each stem bears 1 to 3 leathery leaves from the top. Inflorescences typically of 1 to few flowers arise from sheathed leaf bases. Species with 2 leaves (e.g., *C. bicolor*) belong to the so-called bifoliate group; their flowers are smaller and fleshier, their inflorescences yield up to 4 dozen blossoms. **Similar:** See *Brassavola*. *Guarianthe* inflorescence bears more flowers; columns are small, stems are softer; Mexico to Venezuela only. *Laelia* sepals are narrower, lip less showy; little range overlap. *Rhyncholaelia* flowers are green with narrow sepals; Mexico and northern Central America only (where no *Cattleya* species occur).

Distribution and Diversity: Some 150 epiphytic and rock-dwelling species thrive in low-elevation wet forest. They range from Costa Rica and Trinidad to Brazil and Uruguay. This genus has been expanded to encompass two additional, distinctive groups: rock-dwelling (rupicolous) species in Brazil (e.g., *C. milleri*), formerly placed in *Laelia* and typically bearing orange, yellow, and purple flowers; and small Brazilian species (e.g., *C. coccinea*), formerly known as *Sophronitis* (now defunct), often with brilliant orange-red flowers. Furthermore, bifoliate species distributed from Mexico to Venezuela and Trinidad and once classified as *Cattleya* now have been moved to *Guarianthe* (e.g., *G. aurantiaca*).

Cattleya maxima, a unifoliate species
Andy Phillips

Uncommon form of *C. trianae*, another unifoliate species
Andy Phillips

Ecology and History: Most species are pollinated by large bees that collect nectar or floral scents. Upon landing, their weight depresses the lip, but as the pollinator enters the flower the lip swings upward, trapping the insect against the column. Bees often visit in the morning, when sweet fragrances are most potent. Many species are considered general food mimics, in that they assume a tubular form widely associated with pollinator rewards; however, as they offer no nectar, they must be visited by naive insects that have not yet learned the deception. Meanwhile, *C. violacea* is occupied by crab spiders that mimic floral colors to ambush visiting insects. The genus is named for Englishman William Cattley (1788–1835), the first to raise orchids in greenhouses. Orchids previously had been cultivated outdoors in a climate fatally inhospitable to warm weather plants. One nobleman savagely branded England "the graveyard of tropical orchids." An oft-repeated legend states that the original specimens (*C. labiata*) had been included inadvertently in mossy packing material from Rio de Janeiro by William Swanson. In truth, Swanson, an accomplished botanist, was well aware of the beautiful orchids he had collected, albeit in Pernambuco, miles north of Rio. The allure of *Cattleya* blossoms largely sparked England's fascination for orchids, and the subsequent American craze for orchid corsages. Two countries claim *Cattleya* species as their national flowers: Venezuela (*C. mossiae*) and Colombia (*C. trianae*), the latter because the species' yellow, blue, and red lip matches the national flag.

Cattleya guttata in habitat, a bifoliate species
Ron Kaufmann

C. bicolor, a bifoliate species from Brazil
Ron Kaufmann

Cattleya coccinea, a red-flowering, former *Sophronitis*
species
Andy Phillips

C. caulescens, a rock-dwelling, former
Laelia species
Ron Kaufmann

C. purpurata, example of flowers with deep
purple lips
Ron Kaufmann

C. hoehnei
Ron Kaufmann

Semi-alba form of *C. walkeriana*, which typically
is entirely purple
Ron Kaufmann

C. dowiana, showing color range within genus
Andy Phillips

Chondrorhyncha

Description: A funnel-shaped lip and rolled, backward-swept sepals characterize *Chondrorhyncha* ("kon-dro-RIN-ka"). **Flower:** Yellow or green (rarely white) blossoms project a somewhat concave dorsal sepal. Lateral sepals curl back, the sides rolled inward to form variously open cylinders; petals flex forward. The large funnel-shaped lip, typically white with purple markings, wraps around the column; its outer edge often is frilly. The column is short and snout-shaped (*Chondrorhyncha* refers to the Greek for "cartilage nose"). Midway down the lip reposes a flattened hump (callus) with forward-pointing teeth. **Plant:** Thin, veined leaves with a central crease are arranged in a loose fan. Inflorescences bear a single flower. **Similar:** *Stenia* lacks rolled sepals. *Cochleanthes* and *Warczewiczella* callus is near lip base, bearing parallel ridges.

Distribution and Diversity: Nearly 30 epiphytic species range from Venezuela to Bolivia, favoring wet, low- to middle-elevation forests. Based on genetic evidence the genus may be split into multiple smaller genera, including *Stenotyla* and *Chondroscaphe*. The latter, ranging from Costa Rica to the Andes, are large plants with flowers bearing 2 teeth astride the column tip, and a second callus.

Ecology and History: Several *Chondrorhyncha* species are pollinated by euglossine bees (*Eulaema*). Females visit flowers seeking nectar, attracted by the light, sweet fragrance. Mistaking the rolled sepals for nectar spurs, they probe deeply into the empty tubes (a form of deceit pollination). Eventually exploring inside the lip's funnel the bee contacts the column, and pollinia are glued to the back of her head. *Chondrorhyncha lankesteriana* commemorates Costa Rica's celebrated Lankester Botanical Garden (now part of the University of Costa Rica), founded by British naturalist Charles Lankester (1879–1969). Leaving Southampton for a Costa Rican coffee plantation, "Don Carlos" arrived "at the right place at the right time to join into the active biological exploration of . . . perhaps the most exciting place biologically on our continent" (Williams 1969: 860). He developed a productive friendship with Oakes Ames, supplying the American orchid expert with specimens of more than 100 new species collected on his farm. The orchid garden he established there was the genesis of today's world-famous botanical garden.

C. hirtzii with blunt column and toothed callus
Andy Phillips

Chondrorhyncha velastiguii showing rolled sepals
Franco Pupulin

Chondrorhyncha (*Chondroscaphe*) *embreei*
Ron Kaufmann

C. (*Chondroscaphe*) *amabilis*
Valérie Léonard

C. (*Stenotyla*) *picta*
Franco Pupulin

C. (*Stenotyla*) *lendyana*
Franco Pupulin

Chysis

Description: Dangling pseudobulbs and arresting yellow or white flowers with prominent ridges on the lip are distinctive of *Chysis*. **Flower:** Firm, waxy flowers are large and showy, ranging from bright yellow or orange to brownish purple, occasionally pink or creamy; the inner region usually is more pale. Petals are slightly narrower than sepals. The 3-part lip, yellow or white with purplish-brown spots and crimson stripes, flanks the column with two side lobes; the front portion is notched and deeply ruffled. From the base of the lip rise several parallel, fleshy ridges below the short, curved, and broadly winged column. Most species are pleasantly fragrant, particularly in early morning. **Plant:** Clustered, jointed pseudobulbs are cigar-shaped, with dry, leaf-like sheaths at the base, and often hang downward. Younger pseudobulbs produce several pleated leaves from the tip that dry and are shed with age. The inflorescence emerges from the leaf bases bearing several flowers.

Distribution and Diversity: Approximately 10 epiphytic or rock-dwelling species embellish low-elevation moist forest. They range from Mexico to Peru.

Ecology and History: The pendulous form of *Chysis* plants occurs in many orchids from Asia but is highly uncommon in tropical America. Also nearly unique is the habit of self-fertilization in several species (including *C. aurea*): by the time the flower begins to open, the 8 pollinia have dissolved into a single, lumpy mass. The genus was named for this curious behavior, using the Greek word for "melting." These melted pollinia slide downward onto the stigmatic area and self-pollination is completed, often before flowers have opened fully; fertilized flowers wilt swiftly thereafter. Some species do have quite long-lasting blossoms (e.g., *C. bractescens*). Many *Chysis* flowers, despite bright colors and sweet scents, are strongly inclined toward self-fertilization, which is curious because orchid evolution has been driven by the benefits of attracting insects that deliver genetic material from other plants. Without pollinators the gene pool is constricted and adaptation to changing conditions is slowed, possibly fatally. Typically such an extreme solution is adopted only when pollinators are rare or unreliable (see *Cyrtopodium, Encyclia,* and *Stelis*).

Chysis bractescens
Andy Phillips

C. laevis foliage and hanging, cigar-shaped pseudobulbs
Ron Kaufmann

Cischweinfia

Description: *Cischweinfia* ("si-SHWINE-fee-ah") orchids are distinguished by apple-green, flattened pseudobulbs, tubular flowers, and a hooded column. **Flower:** Small (<1.5" [3.8 cm]) flowers bear green or greenish-yellow sepals and petals, sometimes heavily marked with reddish brown. The outer half of the lip is flared and comes to a point; the inner half wraps tunnel-like around the column and is fused to it at the base. Primarily white, the lip often displays yellow and purple lines within its hairy throat. The short, prominently hooded column has twin arms beneath the stigmatic area. **Plant:** Smooth, flattened, bright green pseudobulbs grow in clusters, each with several leaf-bearing sheaths at the bottom. Leaves are narrow and somewhat leathery. A multi-flowered inflorescence emerges from the pseudobulb base. **Similar:** *Trichopilia* flowers are much larger, column lacks arms.

Distribution and Diversity: Approximately 10 species grow as miniature epiphytes in very wet, low- to middle-altitude forests, from Costa Rica to Bolivia.

Ecology and History: Male and female euglossine bees, searching for nectar within the rolled lip, are known to pollinate *C. dasyandra*. Beyond this single species little is known about pollination in this genus. *Cischweinfia* is named after Charles Schweinfurth, author of the comprehensive, 3-volume *Orchids of Peru*. To understand the genus name requires familiarity with botanical naming conventions. Every plant is known by its genus, species, and the name of the person who first identified it, as well as authors of subsequent taxonomic revisions. For example, the full name of the aforementioned orchid is *Cischweinfia dasyandra* (Rchb.f.) Dressler & N.H. Williams; parentheses enclose the abbreviation of its original classifier, Heinrich Gustav Reichenbach. Vast numbers of 19th-century species were described by Reichenbach and well-known taxonomists John Lindley (Lindl.), and James Bateman (Batem.). Orchids from the 20th century commonly bear names of the subsequent generation of great orchidologists, including Rudolf Schlechter (Schltr.), Oakes Ames (Ames), and Schweinfurth. As his abbreviation Schweinfurth selected "C.Schweinf," applying it to *Bletia amabilis* (C.Schweinf.), and many more. To commemorate his years of service to the world of orchids, the present genus was named after his abbreviation, rather than his full name, by transforming "C.Schweinf" into *Cischweinfia*.

C. parva showing hooded column
Andy Phillips

Cischweinfia platychila
Andy Phillips

C. dasyandra showing pseudobulbs
Andy Phillips

Cochleanthes

Description: Arresting pale flowers feature a rounded lip with radiating markings resembling a shell; *Cochleanthes* takes its name from the Greek words for "shell" and "flower." **Flower:** Cream-colored or greenish sepals and petals are spreading; lateral sepals are united to the column base. Marked by purple radiating stripes, the broad, undulating lip is rounded at the base and shallowly concave to convex. A fleshy, crescent-shaped ridge crosses its base, deeply scored with parallel grooves. A blunt keel projects from the thick column's underside. **Plant:** Bright green, soft-textured leaves grow in fan-like sprays, often densely clustered and obscuring the flowers. Pseudobulbs are absent. An inflorescence usually carrying a single flower arises from the leaf base. **Similar:** See *Chondrorhyncha*. *Kefersteinia* column underside bears a pointed tooth. *Pescatoria* and *Huntleya* lips are convex at midpoint; *Pescatoria* and *Stenia* columns lack blunt keel. *Warczewiczella* flower is throat-shaped, basal lobes enfold column.

Distribution and Diversity: Just 2 species of *Cochleanthes*, as presently understood, grow epiphytically in low- to middle-elevation montane forests and scrub. The genus ranges from Mexico and the Caribbean to northern South America.

Ecology and History: Euglossine bees (*Euglossa* and *Eulaema*) are the primary visitors to *Cochleanthes* flowers. Scratching the lip to collect odors, they crawl inward to the space beneath the column base. Clambering over the curved ridge on the lip, they pick up pollinia on the backs of their heads, or deposit it on the stigmatic region while backing out. Large-bodied male bees are attracted by the strong perfume of these long-lasting flowers (e.g., *C. aromatica*), which they then utilize in mating displays. Plants in *Cochleanthes* usually occur on tree trunks and are adapted to the dim light and high humidity found there. This preference is in contrast to many other orchids that are partial to either the lofty and sun-splashed canopy or the delicate perches on outer twigs (see *Erycina*). Specialization in a particular microhabitat is known as *niche diversification*, and it allows many species to coexist in a single environment. Such fine subdivision of habitat is recognized as a principal cause for the extreme diversity of orchids in the tropics.

Cochleanthes flabelliformis
Andy Phillips

C. aromatica with furrowed ridge at lip base
Andy Phillips

Comparettia

Description: The small plants of *Comparettia* bear delicate flowers with sepals forming a curled nectar spur. **Flower:** Purple, orange, and greenish-yellow colors are dominant. Lateral sepals sweep back to form a sheath around twin nectar-producing horns composed of the basal lobes of the lip; externally the sheath appears as a single tube. While many species produce flowers that open only partially, others have an erect dorsal sepal and forward-pointing petals. In the latter group, the lip is broadly fan-shaped with a deep central notch and a pattern of lines or spots leading to the base; the notch may be lacking in the former group. **Plant:** Small pseudobulbs are flattened ovals or cylindrical, each bearing a relatively large, fleshy leaf. The wiry inflorescence yields several to many downward-nodding flowers. **Similar:** *Rodriguezia* lip is small and narrow, column has twin arms. *Ionopsis* lacks distinct horns.

Distribution and Diversity: Expanded to include former *Scelochilus* species, *Comparettia* comprises more than 70 species of epiphytes inhabiting a wide range of altitudes. The genus is found from Mexico and the Caribbean through the Andes to Bolivia and Brazil.

Ecology and History: Flowers with nectar spurs evolved to attract long-tongued pollinators. Many species (e.g., *C. macroplectron*) show strong adaptations to luring butterflies: colors are bright, the lip provides a large landing platform, and markings serve as nectar guides. Others are specialized to attract hummingbirds, relying on bright reddish-orange colors (e.g., *C. ignea*) or special tissue humps (e.g., *C. coccinea*) that force the bird's beak upward as it feeds, pressing its forehead against the column to effect fertilization. Often the emphasis on hummingbird pollination appears in plants favoring higher elevations, where fewer insects abound (e.g., *C. speciosa*). Nectar production strategies vary. Some species produce copious rewards, attracting greater numbers of pollinators; but often such species are revisited by thirsty hummingbirds, leading to undesirable self-fertilization. *Comparettia falcata* flowers, however, produce a one-time-only secretion of low-sugar nectar: enough of a reward to attract hummingbirds but not enough for them to visit more than once.

Comparettia macroplectron
Ron Parsons

C. ignea, hummingbird pollinated
Ron Kaufmann

Comparettia

Comparettia macroplectron showing nectar spur
Joe Meisel

C. cf. *langlassei* (formerly *Scelochilus*)
Ron Kaufmann

Coryanthes

Coryanthes mastersiana
Andy Phillips

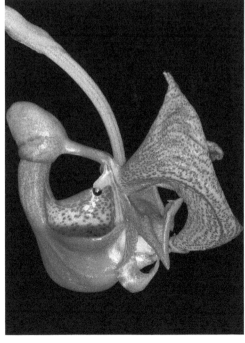

C. mastersiana showing faucet glands; see next page
Andy Phillips

Coryanthes

Description: The Bat or Bucket Orchids of *Coryanthes* are distinguished by back-swept sepals and a complex lip resembling a helmet mounted above a bucket. **Flower:** Yellow with reddish-brown spots or cream-colored, the intricate flowers last only 3 to 4 days, changing rapidly as they mature and wilt. Lateral sepals initially flare outward, then fold back like a bat's wings frozen in the upstroke. A small dorsal sepal bends down, while tiny petals are nearly undetectable. The bizarre lip combines a polished helmet-shaped bulge (*Coryanthes* means "helmet flower"), a smooth central region resembling a water slide, and a deep bucket formed by enlarged, inward-curling side lobes. Paired faucet-shaped glands on the stocky column secrete a clear fluid that drips into the bucket. **Plant:** Clustered pseudobulbs are grooved or fluted, conical to spindle-shaped, each producing 2 to 3 soft, pleated leaves. A hanging inflorescence emerges from the pseudobulb base.

Distribution and Diversity: Approximately 50 epiphytic species favor low-elevation forest from Mexico to Brazil. They grow in ant gardens, which provide rich soil and pest protection, or by wasp nests, so they should be approached cautiously.

Ecology and History: Darwin described how "humble bees" battle for access to the helmet part of the lip, scrabbling at its surface to collect what we now know are fragrance molecules. The scents are later used to attract females for reproduction. Male bees routinely slip down the slide, suffering an "involuntary bath" in the bucket (Darwin 1877: 175). The only exit is through a tunnel past a firm hump that wedges the bee against the column, where pollinia are glued to his back. The tunnel and hump are shaped to favor bees of a particular size, reducing the risk of hybridization by smaller or larger bees. Once pollinated, flowers disintegrate into mush by means of a self-digesting enzyme. Brazil nut trees also are pollinated by bees (*Eulaema*) that collect perfume from *Coryanthes vasquezii* flowers growing in intact rainforest. In disturbed habitats, or where Brazil nut trees are grown in plantations, orchid flowers are scarce and males cannot find fragrance to attract females. Without reproduction, bee populations decline and the unpollinated trees fail to produce these delicious and economically important nuts.

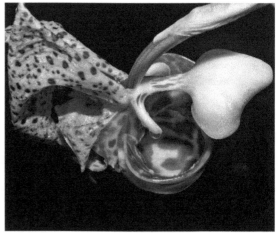

Coryanthes cf. *speciosa*; and see previous page
AWZ Orchids

C. bruchmuelleri flower from above, showing bucket lip
AWZ Orchids

Cranichis

Description: Named after a Greek word for "helmet" (*kranos*), *Cranichis* flowers are distinguished by the helmet-shaped lip held above the column. **Flower:** Small blossoms (1/4–1/2" [0.6–1.3 cm]) are white, pale green, or pink, and delicate in texture. Fine hairs often fringe the petal edges. Flowers are upside down (non-resupinate), with the relatively large white lip presented uppermost. Also fringed, the lip is marked with intriguing patterns of bars, spots, or raised veins. The short column is somewhat pointed. **Plant:** These small herbs have fleshy roots and pliable leaves that are softly pleated and notably stalked. The erect inflorescence is densely packed with up to 50 flowers. **Similar:** *Ponthieva* petals are fused to the column.

Distribution and Diversity: Entirely terrestrial, more than 50 species of *Cranichis* inhabit wet forest across a wide elevational range. Often overlooked, they can be found from Florida, the Caribbean, and Mexico to Brazil and Argentina.

Ecology and History: The pollinators of these small orchids are presumed to be sweat bees (halictids). One of the few groups of bees active only at dawn and dusk, they typically are dark and metallic in color. An aggravating nuisance to people, sweat bees tirelessly fling themselves into open eyes, nostrils, mouths, and any other source of moisture. *Cranichis muscosa*, a common terrestrial in Puerto Rico, achieves its broad distribution through self-pollination rather than reliance on bees. As the flower matures, the anther cap falls off and the pollinia literally drop onto the stigmatic area of the column, accomplishing fertilization. Another group of species in Ecuador shows an intriguing pattern of hybridization. *Cranichis ciliata* prefers the deep shade of closed forest, while *C. lehmanniana* favors open, damp, and sunny meadows. Deforestation eliminated much of the former species' habitat, which was colonized by shrubs that cast a moderate amount of shade. Both orchid species could survive under these bushes, and intermediate-looking hybrids began to appear. Some of those hybrids began to interbreed back with the remaining *ciliata* population, a process known as *introgression*. This transfer of *lehmanniana* genes into the forest-loving orchids underscores the role that intact habitat plays in maintaining separate species.

Cranichis muscosa Valérie Léonard	*C. ciliata* showing fringed petals Joe Meisel	*C. ciliata* showing leaf (lower right) Joe Meisel

Cryptocentrum

Description: The dull-colored flowers of *Cryptocentrum*, viewed from the front, resemble a starfish. **Flower:** Narrow sepals and petals are pale green, tan, or brown, rarely purple, and the sepals are twice as long as the petals. A tiny, tongue-shaped lip is joined to the base of the column. Behind the flower, a long nectar spur is concealed by leafy bracts; the genus name reflects the Greek words for "hidden spur." **Plant:** Small plants with erect stems lack pseudobulbs, except *C. pseudobulbosum* and *C. roseans*. Flat, fleshy, narrow leaves are arranged on opposing sides of the stem, or occasionally in a spiral. A wiry inflorescence bearing 1 flower arises from the stem base. **Similar:** *Maxillaria* flowers lack a spur.

Distribution and Diversity: Some 20 *Cryptocentrum* species grow epiphytically in low- to high-altitude wet forests. They are found from Nicaragua to Bolivia.

Ecology and History: Many species emit a sweet, nocturnal fragrance. The flowers' tubular spur and dull colors perfectly exemplify the moth-pollination syndrome (see *Brassavola*). Moths have poor color vision in the dim light of night, but an excellent sense of smell. Their preferred flowers do not waste energy infusing blossoms with bright colors, relying instead on perfume to attract their nocturnal visitors. The relationship between the lengths of nectar spurs and pollinator tongues has been studied since Darwin's day (again, see *Brassavola*). Modern manipulative experiments have cast light on the pressures driving the evolution of spur length. Pollinators are known to insert their tongue, or proboscis, only as far as is needed to reach nectar. Artificially shortening a spur causes the plant's reproductive success to plummet because pollinators can sip nectar without pressing their heads against the column and its pollinia. Conversely, visitors to flowers with overly long spurs cannot reach the nectar; they ignore such blossoms, again causing fertilization rates to drop. Between these two extremes, minute variations in nectar spur length can reinforce flower-pollinator pairing and enhance reproductive isolation: pollinia are picked up only by pollinators with the perfect tongue length, who carry those pollinia only to other flowers of the same species.

Cryptocentrum cf. *calcaratum* in habitat
Ron Kaufmann

C. standleyi showing foliage and hidden nectar spur
Ecuagenera

Cuitlauzina

Description: The distinctive, white to pink flowers of *Cuitlauzina* are contrastingly marked with vivid yellow on the base of the lip. **Flower:** Sturdy, long-lived flowers can be strongly fragrant, the perfume reminiscent of citrus. Sepals often are narrower than petals; both are brilliant white with occasional pink blush. The outer portion of the white lip usually is wide and notched, the bright yellow base narrowing sharply like a racquet handle. In most species the lip is uppermost (non-resupinate), but others exhibit resupinate flowers. Short and upright, the column bears twin, outspread wings. **Plant:** Densely clustered pseudobulbs are glossy, ovoid, and flattened, yielding 2 leaves from the tip; an inflorescence, hanging or erect, with multiple flowers emerges from the base of the youngest pseudobulb or from new growth. Leaves are strap-shaped with a central crease, and folded at the base. **Similar:** *Oncidium* flowers exhibit a fleshy, warty mass on the lip base.

Distribution and Diversity: Some 10 species dwell as epiphytes and rock dwellers in low-elevation pine and oak forests, ranging from Mexico to Colombia.

Ecology and History: *Cuitlauzina* species have bounced between several genera, but genetic analysis supports an independent genus. The name of this largely Mexican group honors the Aztec nobleman Cuitlahuatzín, a noted designer of early public gardens. He survived just 80 days as governor of Tenochtitlan before succumbing in 1520 to smallpox introduced by Spanish conquistadors. *Cuitlauzina pendula* (a resupinate species), found only in southwestern Mexico, puts forth gorgeous blossoms avidly sought by collectors. Sadly, this species is now endangered because of overharvesting and habitat loss, along with some 15% of that country's 1200 orchid species. Confronted with the failure to protect this orchid, Mexican researchers have resorted to artificially reproducing *C. pendula* in greenhouses. Seeds from botanical gardens are germinated on a growth medium in sterile jars, a feat thought impossible until 1917, when a Cornell University scientist circumvented their dependence on symbiotic fungi (see *Sobralia*). Lewis Knudson (1884–1958) invented a revolutionary medium that provides all nutrients necessary to developing seeds, without symbiotic fungi. Without Knudson, propagation systems that manufacture millions of commercial plants annually, and conservation programs like the one in Mexico, would be impossible.

Cuitlauzina convallarioides, a lip-uppermost species
Andy Phillips

C. pendula, with resupinate flowers
Ron Kaufmann

C. pulchella inflorescence of non-resupinate flowers
Joe Meisel

Cyclopogon

Description: Terrestrial plants with stalked leaves and tubular flowers that lack visible nectar spurs are emblematic of *Cyclopogon*. **Flower:** Sepals and petals of the scented flowers are partly fused, forming a tube; they vary from cream colored to pale green to olive brown. The name derives from the Greek for "circular beard," a reference to the hairy outer surface of the sepals. The white lip flares beyond the floral tube, its enclosed base bearing a pair of hook-shaped lobes astride a small column. **Plant:** Fleshy roots give rise to a whorl of soft green leaves, often with silvery stripes, borne on stalks that may be quite long. Leaves typically wilt before or during flowering. An upright inflorescence, punctuated with small bracts, arises from the leaf bases. **Similar:** *Eltroplectris* lip base lacks lobes. *Pelexia* flowers bear a humped spur below. *Sarcoglottis* flowers have long, leafy bracts beneath them.

Distribution and Diversity: Approximately 80 terrestrial, or rarely epiphytic, species are adapted to a wide range of elevation, moisture, and habitat type. They occur in Florida, the Caribbean, and from Mexico to Argentina and Uruguay.

Ecology and History: *Cyclopogon elatus* is visited by sweat bees (halictids; see *Cranichis*) that extend their long mouthparts deep into the flowers, probing for a hidden repository of nectar. The bee presses its head against the column in the process, and pollinia are adhered to its upper lip, the labrum. Flowers attract the bees with a curious scent that contains the compound DMNT, known as a *wound volatile*, a chemical that nearly all plants on earth emit in response to herbivore damage. Wound volatiles attract predatory insects that attack the leaf-munching herbivores. The origin of floral fragrances may lie in modifications of DMNT to attract insects that pollinate, rather than predate. Studies of *C. cranichoides* reveal a tense balance between reproduction and nutrient availability in these terrestrial orchids. Small plants bear few leaves, and only large plants are able to flower. Furthermore, the bigger the plant the more blossoms are produced, attracting a greater number of pollinators. Some plants do not regain leaves for 1 to 2 years after flowering, surviving underground on decaying organic matter until they accumulate sufficient energy to bloom again (see *Govenia*).

Cyclopogon lindleyanus
Joe Meisel

C. iguapensis showing
leaves and inflorescence
Ron Kaufmann

C. elatus leaves
Ron Kaufmann

Cycnoches

Description: *Cycnoches* flowers are easily recognized by the long, elegant column to which they owe their common name of Swan Orchid. **Flower:** Fragrant, large flowers (to 7" [17.8 cm]) have green, yellow, or brown petals and sepals, and are presented upside down (non-resupinate), with lip uppermost. All species have unisexual blossoms, with male and female flowers quite different. Male flowers are thinly textured, often colorful or patterned, with the characteristic upward-swooping column. Female flowers are more robust and less colorful, their column shorter and tipped with a notched club. The lip, larger in female flowers, has a chin-like fleshy ridge (the callus) near the base. **Plant:** Tall (to 18" [46 cm]), cylindrical pseudobulbs with several horizontal seams often bear papery bracts. Soft, broad leaves are pleated and deciduous. Inflorescences, bearing flowers of a single gender only (5–40 male flowers, or 3–5 females), arise from seams near the apex of the pseudobulb. **Similar:** See *Catasetum. Polycycnis* pseudobulbs are tiny.

Distribution and Diversity: More than 30 species of *Cycnoches* occur in low-elevation humid forest from southern Mexico to northern Brazil.

Ecology and History: Unisexual flowers, an extreme adaptation to reducing self-pollination, are shared only by *Catasetum* and *Mormodes*. Uncommon female flowers are produced only when sunlight, humidity, and nutrients are optimal. Male euglossine bees collect fragrances from the callus of both flower types. A bee begins on the upside-down male lip, unsteadily clutching its enlarged, slippery callus; occasional missteps cause his flailing abdomen to contact the column. In some species (e.g., *C. egertonianum*) the lip is hinged, sagging under the bee's weight. When the column is touched, spring-loaded pollinia are catapulted onto the abdomen (see *Catasetum* for details). The same bee subsequently visits a female flower and departs by dropping upside-down from its lip and rolling over in mid-flight. During this maneuver his abdomen drags across the notched column tip, snagging the pollinia. Male flowers are short lived and wilt within hours of discharging pollinia to deter return visitors. Seed capsules take a year to develop, requiring copious energy found only in the ideal conditions that support female flowers.

Cycnoches barthiorum female flowers
Fred Clarke

C. cooperi male flowers
Fred Clarke

C. herrenhusanum male flowers
Fred Clarke

C. egertonianum male flowers
Ron Parsons

Cyrtochilum

Description: Rambling, twining inflorescences, sometimes more than 10 feet (3 m) long, and showy flowers distinguish *Cyrtochilum*. **Flower:** Small to large flowers usually are yellow with brown speckling or wash, less commonly purple on white. Sepals are broad and spreading, with undulating edges; petals are slightly smaller but similarly colored. The relatively short lip is arrow-shaped, its tip elongated and curled down (*Cyrtochilum* is from the Greek for "curved foot"). On the lip's base lies a warty or toothed mass of tissue (callus). **Plant:** Pseudobulbs are variable in shape, usually widely spaced, and are largely covered by long sheaths and old leaf bases. Each produces several opposing pairs of long, narrow leaves with a pleated or ribbed surface. **Similar:** *Oncidium* inflorescence is not long and twining; pseudobulbs are clustered, less obscured.

Distribution and Diversity: More than 130 epiphytic and terrestrial species brighten middle- to high-altitude cool forests. They range through the Andes from Venezuela to Peru, with one species in the Caribbean.

Ecology and History: *Cyrtochilum* shares many ecological characteristics with *Oncidium*, from which it is separated on the basis of molecular data (see *Trisetella*). Flowers of most species are pollinated by a variety of bees, some seeking oil rewards. Bright orange displays of *C. retusum*, however, are visited by hummingbirds; the unusual, dark color of the pollinia likely evolved to avoid the bird's notice once affixed to its bill (see *Elleanthus*). This genus includes many breathtaking species popular with collectors. *Cyrtochilum flexuosum* was probably the first orchid exhibited in the United States, in 1837. Such plants were assiduously pursued across the mountains of tropical America by orchid hunters employed by European collectors. Innumerable dangers and hardships faced these intrepid men, not least of which was attack or murder by rivals in the employ of other collectors and greenhouses. Practically all contracted malaria or yellow fever, and many died from their illnesses. *Cyrtochilum weirii* commemorates John Weir, an orchid hunter dispatched to Brazil who was stricken with an unknown disease rendering him "paralysed in all his limbs from the neck downwards" (Anonymous 1898: 301).

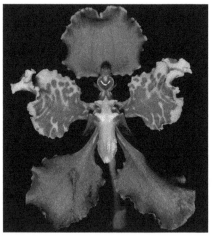

Cyrtochilum serratum
Ron Kaufmann

C. ionodon showing warty callus
Andy Phillips

Cyrtochilum macranthum
Andy Phillips

C. weirii
Ron Parsons

C. retusum, hummingbird-pollinated
Ron Parsons

C. pastasae
Andy Phillips

Cyrtopodium

Description: *Cyrtopodium*, from the Greek words meaning "curved foot" because of the distinctly bent column, are also called the Cigar Orchids or Cow's Horn Orchids for their elongated, spindle-shaped pseudobulbs. **Flower:** Sepals often curl forward and have undulating edges, while petals are smooth and curl back slightly. Both typically are yellow with reddish-brown markings. Side lobes of the lip nearly enclose the column; between them lays a raised pad (callus) composed of distinctive, fleshy tubercles resembling a field of tiny warts. **Plant:** Tall, cigar-shaped pseudobulbs may reach 4 feet (1.2 m) in height. They retain sheathing leaf bases that usually terminate in sharp, spine-like tips. A robust, branching inflorescence (to 48" [1.2 m] tall) bears many flowers. Leaf-like bracts the same size and color as the sepals arise from each joint of the inflorescence. **Similar:** *Mormodes* and *Galeandra* have dissimilar flowers, typically on shorter, unbranched inflorescences.

Distribution and Diversity: Nearly 50 terrestrial and epiphytic species are widespread, occurring in Florida and the Caribbean, and from Mexico to Argentina.

Ecology and History: Many *Cyrtopodium* orchids grow in seasonal climates and drop their leaves during rainless months; the spines that remain protect pseudobulbs against hungry herbivores. The genus is pollinated primarily by large bees. Lacking nectar, the flowers are long-lived and often strongly fragrant. The bees are steered beneath the column by the lip's side lobes, ensuring pollination. Several species mimic flowers of other plants: *C. punctatum* imitates the oil-producing flowers of a shrub (*Byrsonima*, or Locust-Berry) and is pollinated by *Centris* bees that mistakenly visit the orchid in search of oils (see *Gomesa*). Orchids located near dense *Byrsonima* populations show much higher rates of pollination, prompting some to suggest planting these shrubs near endangered *Cyrtopodium* orchids to boost their reproduction. Another species from Brazil, *C. flavum*, mimics a related oil-producing shrub and a nectar-providing bean bush (*Crotalaria*, or Rattlepod), luring both plants' pollinators. These insects are grounded during the region's abundant rainstorms, but the orchids can exploit rain-assisted self-pollination instead: during heavy rainfall, droplets accumulate on the stigmatic zone of the column, contact the pollinia, and draw pollen onto the reproductive zone.

C. punctatum, with bent column
Ron Parsons

C. flavum showing leafy bracts
Andy Phillips

Cyrtopodium plants in dry habitat
Ron Kaufmann

Dichaea

Description: *Dichaea* can be recognized by the overlapping leaves, on opposite sides of the stem, twisted so they all face in the same direction. **Flower:** White, yellow, or purple sepals and petals, typically with dense violet spots, are pointed and spreading, giving a star-shaped appearance. The lip often is 3-lobed and anchor-shaped, and attached narrowly to the base of the short column. **Plant:** Scrambling or pendent stems lack pseudobulbs. Two-ranked leaves are small, stiff, and distinctly folded in half, giving the genus its name (Greek for "twofold"). Inflorescences bearing a single flower arise from leaf bases. **Similar:** *Lockhartia* leaves are flattened laterally (hiding the central fold), thicker, and more closely packed. *Fernandezia* plants are tiny (to 6" [15 cm]), inflorescences bear multiple flowers (but if single, bright reddish orange).

Distribution and Diversity: Approximately 120 species of *Dichaea* inhabit low- to middle-elevation wet forests where they grow as epiphytes. The genus ranges from Cuba and Mexico to Peru and Brazil, with peak diversity in the Andes.

Ecology and History: *Dichaea potamophila* plants exhibit synchronous flowering, daytime release of a slight musty scent, and nocturnal closing of flowers. These blooms attract large, day-active euglossine bees (*Eulaema*) that collect fragrance compounds from the lip. Another species, *D. schlechteri* (synonymous with *D. similis*), was named for legendary German taxonomist Rudolf Schlechter (1872–1925), who tartly identified the one crucial skill for a career in botany: "Without a good memory it is of no use trying to be a botanist; one had better give it up and be a merchant" (Reinikka 1995: 293). His stated intention was to describe one new species every day of his life. Numerically unsuccessful, he did propose more than a thousand species and is commemorated today by the genus *Rudolfiella*. A third species, probably *D. splitgerberi*, is used by the Kofán people of Colombia to treat eye infections. Their medicines were studied by Harvard ethnobotanist Richard Schultes, who, during marathon expeditions to the Amazon, never carried a firearm, saying, "I do not believe in hostile Indians. All that is required to bring out their gentlemanliness is reciprocal gentlemanliness" (Sequeira 2006: 9).

Dichaea trichocarpa
Andy Phillips

D. hystricina
Andy Phillips

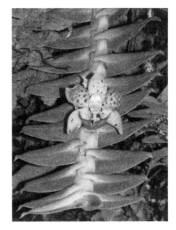

D. longa showing overlapping leaves and central fold
Ron Kaufmann

Dracula

Description: The genus *Dracula* is named not for the Transylvanian plasma connoisseur but rather for the little dragons whose faces the flowers resemble, with tiny petals for eyes and a cupped lip for a snout. **Flower:** All 3 sepals are united for a third of their length, forming (usually) a downward-hanging bell or plate; they are densely hairy and terminate in long tails. Petals are much smaller, often with dark markings, on opposite sides of a blunt, clown nose–like column. The hinged white lip is cupped, like a ladle or shell, and exhibits radiating veins or ridges strongly reminiscent of the underside of a mushroom. **Plant:** Short stems lack pseudobulbs, giving rise to thin leaves with a distinct midrib. **Similar:** *Masdevallia* flowers do not resemble a face; the lip is immobile and lacks ridges.

Distribution and Diversity: Favoring cool conditions, more than 120 species of *Dracula* lurk epiphytically or on rock faces in middle- and upper-altitude forests. The genus can be found from Mexico to Peru.

Ecology and History: These arresting flowers are known in some regions as *cara de mono*, or monkey face (e.g., *Dracula simia*). Mimicry of mushrooms by the lip, however, is what attracts pollinators. Fungus gnats (*Zygothrica*) are tiny flies that mate on top of toadstools, deposit eggs there, and occasionally feed on the mushroom itself or yeasts growing on it. These diminutive flies find mushrooms by visual and olfactory cues, many imitated by *Dracula* flowers: cupped, wrinkled lips resemble the underside of a mushroom cap, blossoms release a smell of decaying vegetation or fungus, and long tails of sepals catch the attention of flying insects (see *Phragmipedium*). Fungus gnats often land on these tails, then climb to the center of the flower. Male flies appear to use the lip as a display arena to attract receptive females, rather than a prospective egg-laying site. During such displays the flies contact the column, pick up and deposit pollinia, and fertilize the flower. Intriguingly, plants flowering near mushrooms favored by these gnats show higher rates of successful reproduction. Thus, conservation of *Dracula* populations depends on protection of such mushrooms, themselves reliant on preservation of humid, mature forest.

Dracula vampira showing the characteristic tails
Andy Phillips

D. wallisii
Andy Phillips

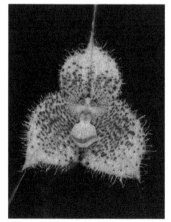
D. diana
Andy Phillips

Dracula simia, the true "monkey face" flower
Ron Kaufmann

D. platycrater showing leaves
Ron Kaufmann

D. brangeri
Andy Phillips

D. ubangina
Andy Phillips

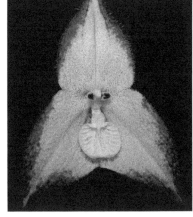

D. chimaera in habitat
Ron Kaufmann

Close-up of *D. cordobae* lip and petals
Joe Meisel

Dresslerella

Description: *Dresslerella* flowers resemble a cross between a Dutch clog and a fuzzy bedroom slipper, while leaves bear short, white hairs. **Flower:** Sepals are covered with hairs, either a short pelt or dense long filaments. Lateral sepals are fused for most of their length, forming a concave shoe housing the column and lip. The dorsal sepal, similarly fuzzy, is erect or lies across the opening. Tiny petals occasionally are visible as jutting antennae, or are hidden within the cavity along with the lip and column. **Plant:** Growing as clumped stems, these small plants lack pseudobulbs. A single leaf arises from each stem, thick, leathery and densely hairy, with a distinct stalk. In a few species, the leaves lie on the ground in a flat rosette. Inflorescences arise from leaf bases, bearing a single flower that often rests on top of the leaf surface. **Similar:** *Myoxanthus*, *Restrepia* leaves and flowers lack dense hairs.

Distribution and Diversity: Some dozen species live as epiphytes or occasionally grow on rocks in humid, middle-elevation forest. *Dresslerella* occurs from Belize and Costa Rica to the Andes of Colombia, Ecuador, and Peru.

Ecology and History: *Dresslerella* is named for respected orchid expert Robert Dressler (see *Trisetella*). Researchers believe, as with most closely related genera (*Dracula*, *Lepanthes*, *Masdevallia*, *Pleurothallis*, and more), that their flowers are pollinated by flies, for which the purplish colors and cupped shape are characteristically attractive. Of greater interest is the overt presence of dense hairs. These small plants grow in mountainous forests that typically experience near-constant cloudiness and regular precipitation. Occasional dry spells occur, however, during which the modest-sized plants risk dehydration. Rather than depending on storage organs, like the pseudobulbs of many other epiphytic orchids, *Dresslerella* plants appear to draw moisture directly from the omnipresent mountain fog. Humidity in the clouds condenses onto the fine hairs of the leaves and flowers as tiny droplets of water. This so-called *cloud stripping* supplies moisture to the plant even when true rain does not fall. Such an approach is characteristic of other diminutive plants, such as mosses and liverworts, but is uncommon in orchids.

Dresslerella caesariata showing
typical leaf-top flower
Joe Meisel

D. lasiocampa
Andy Phillips

D. hirsutissima showing antennae-
like petals
Andy Phillips

Dryadella

Description: The tiny, usually speckled flowers of *Dryadella* peek out from clumped vegetation, earning one species (*D. edwallii*) the nickname of Partridge in the Grass. **Flower:** Minute flowers (1/2" [1.3 cm]) typically are orange to yellow with dense purple spots. Lateral sepal bases are fused to form a slight cup, crossed by a dam-like hump (callus); lateral sepals bend sharply at this callus. All sepal tips end in short tails. Shorter petals are broad and shaped into many-angled geometric forms. The spade-shaped lip is narrowly attached to the base of the slender column. **Plant:** Dwarf plants are composed of tufts of leaves (to 2" [5 cm] long) on short stems. A stubby inflorescence bears a single flower, obscured within the clustered leaves. **Similar:** See *Barbosella*. *Masdevallia* dorsal sepal is fused to lateral sepals. *Pleurothallis* sepals do not bend back at the callus.

Distribution and Diversity: Approximately 50 epiphytic species favor wet forests, or occasionally hillside scrub, across a range of elevations. They occur from Mexico to Brazil and Paraguay.

Ecology and History: Most species emit a foul odor, attracting minute flies that seek rotting vegetation or corpses on which to lay eggs, a syndrome known as *sapromyophily* (see *Bulbophyllum*). The hinged lip is delicately balanced, the weight of the fly tipping it upward, pinning the hapless pollinator against the column. All *Dryadella* species fall convincingly in the diverse category of orchids known as miniatures. A great number of these miniatures are exceptionally popular with hobbyist orchid collectors: their small size allows many species to be reared in a modest terrarium or greenhouse, they often display entrancing little flowers, and a nearly endless diversity of miniature species exist. Indeed, the most diverse genera of orchids are almost exclusively miniatures: *Pleurothallis* (nearly 600 species), *Stelis* (nearly 900), and *Lepanthes* (more than 1000). Although *Dryadella* contains fewer species, they are no less miniature in scale. Many of the species names reflect their diminutive stature. *Dryadella gnoma* refers to squat gnomes, *D. lilliputiana* to Gulliver's tiny captors, *D. minuscula* literally means minuscule, and *D. pusiola* comes from the Latin word for "little girl."

Dryadella lilliputiana
Ron Parsons

D. edwallii flowers peeking from amid its leaves
Andy Phillips

D. zebrina close-up showing flower details
Andy Phillips

Elleanthus

Description: *Elleanthus* plants are distinguished by large, pleated leaves and spearhead- or ball-shaped clusters of densely packed, small flowers. **Flower:** Often dull whitish or yellow in color, occasionally orange or purple, the small flowers are clothed by waxy bracts, which are modified leaves resembling petals. These bracts, their color dominating the inflorescence, are brilliant purple to reddish orange. Sepals and petals are similar, and the lip often is sac-shaped with a hairy or toothed outer edge. **Plant:** Tall, reed-like stems may be simple or branched, and commonly reach 4 feet (1.2 m) in height or more. Ribbed or folded leaves with pointed tips emerge from the stem alternately and in 2 ranks, like the symmetrical sides of a feather. All species lack pseudobulbs. Inflorescences arise from atop the stems and bear open and unopened flowers simultaneously. **Similar:** *Sobralia* inflorescence bears fewer, but larger flowers.

Distribution and Diversity: More than 100 species grow terrestrially, a few epiphytically, in low- to high-altitude forests. They occur in Cuba and Trinidad, and from Mexico to Brazil, reaching their peak abundance in the Andes.

Ecology and History: *Elleanthus* flowers are an excellent example of the hummingbird pollination syndrome (see *Cochlioda* and *Comparettia*). Blossoms typically are tubular in shape, are advertised by bright red or purple bracts, produce nectar and lack scent, and are presented in an open, readily accessible space. Pollination by hummingbirds may explain the unusual dark bluish-gray color of the pollinia of numerous species, such as *E. capitatus*. Such color is rare among orchids, which typically have bright yellow pollinia. One theory holds that vivid yellow pollinia affixed to a hummingbird's blackish beak would cause the bird to fixate on this bright spot and be more likely to remove the pollinia before reaching another flower. The unusual coloration may even spare hummingbirds from becoming cross-eyed. This droll theory is supported by the observation that 2 species not pollinated by hummingbirds, *E. caricoides* and *E. linifolius*, have yellow pollinia. *Elleanthus* inflorescences sometimes are covered in a transparent, jelly-like slime before the flowers open. This mucus likely is a protection against rapacious slugs and snails that could destroy flowers before they are pollinated.

Elleanthus robustus flower close-up
Ron Kaufmann

E. capitatus ball-shaped inflorescence, protective jelly
Ron Kaufmann

Elleanthus robustus spike-shaped inflorescence
Ron Kaufmann

Yellow-flowered *Elleanthus* showing foliage
Ron Kaufmann

E. aurantiacus, hummingbird-pollinated
Andy Phillips

E. purpureus spike-shaped inflorescence
Ron Kaufmann

Eltroplectris

Description: The unusually large, pale flowers of *Eltroplectris* differentiate this genus from other terrestrials. **Flower:** Narrow, lateral sepals flare outward, their bases fused with that of the lip to form a prominent, curving spur. The spur's tube may be joined to the flower stalk, but the tip always is unattached: the genus name refers to the Greek words for "free spur." A hood-like dorsal sepal narrows to an elongated point. The tongue-shaped lip is frilly or notched. Strongly fragrant flowers range in color from white to pink or greenish tan. **Plant:** Broad leaves, bluish to purplish green and often with spotting underneath, narrow somewhat at their bases. Leaves lie near the soil in a rosette or cluster, rarely solitary on a long petiole, and are deciduous. Inflorescences (to 3' [0.9 m] tall) hoist loosely arranged flowers, often opening after the leaves are lost. **Similar:** See *Cyclopogon. Eurystyles* flowers are tightly packed, plants epiphytic. *Pelexia* lip bears small basal lobes. *Sarcoglottis* spur tip is not free.

Distribution and Diversity: A dozen terrestrial species favor lower-elevation humid forest in Florida and the Caribbean, and from Mexico to Paraguay.

Ecology and History: *Eltroplectris* pollination remains unstudied, but the combination of pale flowers, sweet fragrance, and nectar spurs suggests nocturnal visitation by moths. Plants occur in scattered populations, overlooked because of their dark leaves; however, the pale blossoms signal the presence of these elusive terrestrials. *Eltroplectris calcarata*, a species found throughout the American tropics, occurs in southern Florida where it inhabits coastal hammocks: island-like mounds of soil that rise above surrounding swamp waters, supporting red cedars, cabbage palms, and numerous terrestrial plants. Coastal hammocks face a direct threat from global warming, as even a modest rise in sea level could drown their flora. *Eltroplectris calcarata* populations are further ravaged by wild pigs, which were introduced to Florida by Spanish conquistador Hernan DeSoto, who perpetually traveled with enormous, seething herds to feed his forces. In 1923 the Fennell Orchid Company of Homestead, Florida, founded the nation's first orchid reserve, the Orchid Jungle, with paths winding through a small, wooded hammock.

Eltroplectris (aka *Pteroglossa*) *roseoalba*
Franco Pupulin

Encyclia

Description: *Encyclia* can be identified by the Dumbo-like ears framing the column and the garlic-shaped pseudobulbs. **Flower:** Sepals and petals are similar, but flower color and size varies. The lip, rarely fused to the column but usually free, extends large wing-shaped lobes alongside the column base. In some species the wings encircle the column, an arrangement for which the genus is named. The column is thin and long, often flaring laterally. **Plant:** Clustered pseudobulbs resemble Russian onion domes, or chocolate kisses. Each produces several fleshy or stiff leaves of varying shape. Inflorescences arise from on top of the pseudobulbs, bearing many flowers, sometimes reaching 3 or more feet (0.9 m) in length. **Similar:** See *Prosthechea*. *Epidendrum* lacks garlic-shaped pseudobulbs; lip and column are fully fused.

Distribution and Diversity: More than 150 species of *Encyclia* thrive as epiphytes in low-elevation dry forest, from Florida and Mexico to Paraguay.

Ecology and History: Most species are pollinated by large bees, especially carpenter bees. The insect alights on the lip, which droops under its weight, opening a path beneath the column and between the lip's wings. Upon backing out, the bee picks up or deposits pollinia. Many of these species attract pollinators by means of fragrance, perhaps none so intoxicating as *E. phoenicea* flowers that smell mouthwateringly of chocolate. A few species are capable of self-pollination if they occur where traditional pollinators are absent (e.g., *E. bradfordii*, *E. oncidioides*) or at the edge of the pollinator's range. Favoring seasonally dry forests, these orchids show a number of adaptations to surviving in nutrient-poor environments. Species such as *E. cordigera* and *E. alata* form relationships with ants, bribing the insects with nectar in exchange for defense from herbivores. Such ant–plant relationships are significantly more common in dry regions, where scarce food resources demand greater protection. The roots of *E. tampensis* also take up nutrients from the host tree bark, uncommon among orchids. Sadly, the original type species, *E. viridiflora*, from the Atlantic Forest of Brazil, is now extinct due to rampant destruction of this ecosystem (see *Leptotes*).

Encyclia alata showing lip's winged lobes
Andy Phillips

E. phoenicea, the chocolate-scented flower
Andy Phillips

Encyclia spiritusanctensis in habitat
Ron Kaufmann

E. fowliei
Andy Phillips

E. oncidioides, a self-pollinating species
Ron Parsons

E. cordigera, an ant-partnering species
Andy Phillips

E. tampensis
Ron Parsons

E. euosma showing typical pseudobulb shape
Ron Kaufmann

Epidendrum

Description: *Epidendrum* flowers exhibit complete fusion of the column and lip, and typically have erect, cane-like stems. **Flower:** Narrow sepals and petals contrast with a flaring lip. Highly variable in color and size, most flowers are small (to 1" [2.5 cm]). The underside of the column appears welded to the lip, with a tiny slit remaining for pollinators. Many flowers are fragrant. **Plant:** Most species have thin, jointed, bamboo-like stems with narrow leaves; a few produce slender pseudobulbs. Inflorescences arise from the top of stems, bearing 1 to many flowers. **Similar:** See *Barkeria* and *Encyclia*.

Distribution and Diversity: More than 1300 species of *Epidendrum* (derived from the Greek for "on trees") grow epiphytically; also a number of conspicuous species are terrestrial. This abundant, widespread genus ranges from Florida to Argentina.

Ecology and History: Most *Epidendrum* flowers are pollinated by butterflies: they insert their tongue beneath the column to probe for nectar, their head contacting the pollinia in the process. Some bright orange species (e.g., *E. ibaguense*, *E. secundum*) are visited by hummingbirds that similarly insert their bill betwixt column and lip. In southern Ecuador, two color morphs of the latter species do not hybridize because they are served by different species of hummingbirds in northern Ecuador; however, a third hummingbird visits both morphs, and numerous hybrids occur. Thus pollinators can keep (some) populations reproductively isolated. Another species, *E. falcatum*, releases a jasmine-like scent in the morning, attracting moths, but later switches to a butterfly-enticing lily perfume. Several species adopt floral mimicry, including *E. radicans* and *E. ardens*, that closely resemble the flowers of nectar-producing shrubs. Butterflies are deceived into visiting the orchid, although it offers no sugary reward. Some species have developed close associations with ants (e.g., *E. denticulatum*), providing external nectar on buds and flowers to attract ants as protectors against herbivores. Others (e.g., *E. flexuosum*) grow in ant nests and even derive nutrients from their waste. Such intricate associations can thwart conservation plans; endangered *E. ilense* plants can be reared indoors, for example, but perish when reintroduced to the wild.

Epidendrum radicans
Ron Kaufmann

E. secundum, purple color morph
Ron Kaufmann

E. calanthum
Ron Kaufmann

Epidendrum nocturnum, moth-pollinated
Andy Phillips

E. bangii
Ron Kaufmann

E. hugomedinae, example of cupped flower form
Andy Phillips

E. falcatum with deeply lobed lip
Andy Phillips

Typical terrestrial *Epidendrum* leaves and stem
Ron Kaufmann

E. escobarianum, showing alternating leaves with
uncommon banding pattern
Ron Kaufmann

Epidendrum lehmannii
Joe Meisel

E. gnomus showing broad lip and fleshy leaves
Ron Kaufmann

E. ilense
Andy Phillips

Robust inflorescence of *E. bifalce*
Ron Kaufmann

E. fimbriatum in typical, dense
terrestrial habitat
Ron Kaufmann

E. embreei, example of
hanging, orange flowers
Ron Kaufmann

E. cf. *difforme* showing range of forms
in genus
Ron Kaufmann

Epistephium

Description: Bamboo-like stems with shiny, leathery leaves and dramatic purple flowers make *Epistephium* easy to recognize. **Flower:** Large, showy flowers are rich magenta, uncommonly pink or green. Sepals and petals are similar, often faintly striped. The trumpet-shaped lip is rolled around the column, with the outer edge flaring open and ruffled. **Plant:** Tall, cane-like stems grow in clusters, occasionally running along the ground before bending skyward. Stems can be among the tallest of all orchids, sometimes towering more than 15 feet (4.6 m) in height. Variably shaped leaves are leathery and glossy, with distinctive reticulated veins. The inflorescence arises from the stem apex and bears several flowers, opening one at a time from bottom to top. **Similar:** *Sobralia* leaves have a matte surface, with pleats and parallel veins.

Distribution and Diversity: Some 20 terrestrial species inhabit low- to middle-altitude forests, savannas, and wetlands. They occur in Trinidad and Belize, and from northern South America to Argentina.

Ecology and History: *Epistephium* are closely related to *Vanilla*, although the plant and flowers more closely resemble *Sobralia*. Based on their shape, flowers likely are pollinated by the same large euglossine bees as the latter two genera. Once pollinated, flowers develop into capsules packed with tiny seeds that are winged, a rarity among orchids. The parachute-like shape helps the seeds disperse farther by catching gusts of wind. Several species, including *E. parviflorum* and *E. lucidum*, inhabit seasonally dry grasslands where such gusts are frequent and powerful. These species flower during the dry season and fruit as the cyclical rains return, a common feature among dry forest plants: fruiting and flowering are tightly synchronized to the climate, ensuring that seeds drift to the parched ground just when moisture becomes available again. *Epistephium brevicristatum* grows in dense stands along rivers of the Colombian Amazon, where the Kubeo people use it as an antiseptic. Infected wounds are washed with water in which the plant has been boiled, and a powder made from dried, crushed flowers is applied topically. Laboratory research on leaf extracts from *E. sclerophyllum*, another rainforest species, demonstrated significant antimicrobial activity, validating the traditional knowledge of Kubeo healers.

Epistephium duckei showing funnel-shaped lip
AWZ Orchids

E. elatum inflorescence
Franco Pupulin

Eriopsis

Description: The unique flowers of *Eriopsis* are readily identified, their yellow sepals and petals trimmed in red and white, and the lip tip spotted with purple. **Flower:** Broad, waxy sepals and petals are pointed and spreading. A yellowish-orange base color fades attractively to reddish purple at the edges. The wide lip is distinctly 3-lobed: both sides of its base curl around the arching column, while the white- and lavender-splotched tip projects forward like a small spoon. Flowers are quite fragrant, with notes of melon and apricot. **Plant:** Clumped, elongated pseudobulbs are dark green to nearly black, their surface often rough. Overlapping sheaths wrap the base of the pseudobulb, which bears 2 to 3 stiff, leathery leaves from its apex. A distinctive "mini-bulb" develops where the pseudobulb and leaves meet. Upright inflorescences producing numerous flowers arise from the plant's base.

Distribution and Diversity: Five species of *Eriopsis* ornament shaded, humid forests from low to middle elevation, often on rocky or clay-rich slopes. They occur from Belize to Peru and Brazil.

Ecology and History: The lovely blossoms of *E. biloba* attract *Eulaema meriana* bees, drawn to their attractive scent and striking colors. Fragrant flowers are common in dense, shady forests, where perfume travels farther than the eye can see (see *Anguloa*). Thick forests also filter the sun, producing a low-light environment in which dark-leaved plants are more common: red and purple pigments help leaves absorb dim light, and lend them a characteristic blue-black appearance. This relationship may explain the dark-colored pseudobulbs of many *Eriopsis*. The tall, hefty pseudobulbs of *E. sceptrum*, a lowland species found in the Amazon, produce copious sticky mucilage. The Makuna people living along Colombia's Apaporis River boil the orchid to extract this gummy substance, which is capable of absorbing prodigious amounts of moisture. They utilize the mucilage as a treatment for painful sores on gums and mucous membranes. Their local name for the orchid, *wan-oo-ma-ká*, literally means "mouth medicine." Other indigenous groups use the same substance to treat fungal skin infections, a testament to the pharmacological value of the rainforest and the ethnobotanical traditions of indigenous peoples.

Eriopsis biloba
Ron Parsons

E. sceptrum
Andy Phillips

E. biloba dark pseudobulbs and "mini-bulbs"
Ron Kaufmann

Erycina

Description: The vivid yellow flowers of *Erycina* resemble a flamenco dancer in motion and nearly dwarf their relatively small plants. **Flower:** Dominating all other floral parts, the large lip is deeply divided into 3 or 4 lobes. Both the lip base and small petals often are marked with reddish-brown bands. Lateral sepals are tiny and obscured, while the small dorsal sepal curves forward over the column. Convoluted wings hood the end of the column, the tip of which is marked with a red dot. **Plant:** All species are small. Most lack pseudobulbs, growing as a tidy fan of alternating, blade-like leaves. The few species that do have small pseudobulbs bear just 1 or 2 leaves from each. **Similar:** *Oncidium* plants are large with sizeable pseudobulbs; flowers are more evenly proportioned, with sepals and petals not dwarfed by the lip. *Ornithocephalus* plants are small to medium sized, typically with white flowers much smaller than the plant.

Distribution and Diversity: Seven species grow epiphytically in seasonally dry forests, from low to middle elevation. They are found from Mexico and Trinidad to Ecuador.

Ecology and History: *Erycina* belong to a group of plants known as twig epiphytes. These diminutive orchids are capable of clinging to tree trunks, but typically grow on thin branches or even the surfaces of leaves. Such microhabitats are unusual for orchids because they are transitory: leaves are shed, and twigs can snap under even moderate weight or in strong winds. Most orchids are long-lived and take years to reach reproductive age. If they colonize a twig, they risk having their perch fail before they can flower. *Erycina* species, however, have evolved to live life at top speed: they can flower in a year, and typically wither after only 6 years. This short life cycle allows them to specialize on relatively ephemeral substrates with little or no competition from other orchids and to reproduce before they lose their footing. *Erycina crista-galli* even grows abundantly on coffee bushes, perhaps aided by their rapid growth. Coffee plantations, especially those that retain shade trees, provide important biological refuges in agricultural landscapes (see *Scaphyglottis*).

Erycina crista-galli
Ron Kaufmann

E. pusilla and fan of leaves
Joe Meisel

E. crista-galli flower close-up
Ron Kaufmann

Eurystyles

Description: *Eurystyles* can be recognized by the combination of a compact cluster of flowers tipping a fuzzy stalk that hangs from leaves arranged in a rosette. **Flower:** Tiny flowers are tubular, surrounded by conspicuous fringed bracts, and packed into a spiral head. Translucent sepals and petals are white to pink to dull brown, hairy on the exterior, and often with a prominent green midvein. The blunt lip is fused to the lateral sepals at the base. **Plant:** Thin, delicate leaves, often glossy and stalked, emerge in a whorl. A thin, hairy inflorescence descends from the center of the leaves. **Similar:** See *Eltroplectris*.

Distribution and Diversity: Some 20 species of miniature epiphytes occur in a range of elevations and habitats. The genus can be found from Mexico and the Caribbean to Paraguay and Brazil, the last harboring their peak abundance.

Ecology and History: The genus *Eurystyles* was cursed from the beginning with an ill-fitting name. Forged from the Greek words for "broad pillar," the name is an erroneous reference to the column. The delicate tissues of the original *Eurystyles* specimen were badly deformed by the drying process applied to all botanical collections (see *Brachionidium*). The flower column, normally quite elongated and slender in living specimens, was reduced to an ignoble stump, and taxonomist Heinrich Wawra applied the name he thought fit best. Unfortunately, he compounded his error by cataloguing the plant as a type of ginger. It took 60 years before the wayward genus was noticed by orchid experts. At that time an appropriate name might have been *Leptostyles*, or slender pillar, but once inscribed, botanical names cannot be changed. Studies of living specimens reveal that nectar production is common in the genus, suggesting bee or moth pollination. *Eurystyles standleyi* was named, this time accurately, for Paul Standley (1884–1963), an intrepid botanist and explorer of Central American jungles, particularly around the canal zone of Panama. He worked 14 hours a day, smoked ceaselessly, and never married. About the region he enthused, "there certainly are plenty of things here to keep me busy, and as long as I am able to crawl around I can find plenty of field work in Central America that is worth while" (Ossenbach 2009: 200).

Eurystyles cotyledon
Ron Kaufmann

E. cf. *auriculata*
Ron Kaufmann

E. cogniauxii plant with early inflorescences
Ron Kaufmann

Fernandezia

Description: Needle-leafed dwarf plants, sometimes mistaken for moss, with relatively large flowers characterize *Fernandezia*. **Flower:** Attractive blossoms, larger than the leaves, are red or orange, occasionally yellow or pale green. Sepals and petals are similar and project forward like a bell. The large, fan-shaped lip is broad at the tip and deeply concave at the base. An unusual column appears sheathed: its broad lateral wings curl down, clasping the lip base. **Plant:** Diminutive plants (but 8" [20 cm] tall in *F. tica*) lack pseudobulbs; adventitious roots occasionally sprout from the stem. Dark green leaves, sometimes purple tinged below, are fleshy and needle-like, the tips curving upward, the edges frequently toothed. Leaves are arranged on two opposing sides of the stem, their sheathing bases overlapping, but not fully. Short inflorescences arise from the bases of the uppermost leaves, bearing few flowers. **Similar:** see *Dichaea*. *Lockhartia* leaf bases obscure stem.

Distribution and Diversity: *Fernandezia* comprise some 50 species. They favor cold, wet, high-altitude sites from Mexico to Bolivia.

Ecology and History: Plants living at high elevation face cold days, frigid nights, and thin mountain air that rapidly steals water from tissues. In response, high-altitude plants have tiny, thick, tough-skinned leaves, resembling spruce needles: the hard skin resists drying while fleshiness permits water storage. The minute size and plump shape of the leaves also reduce the surface area across which water and heat can be lost. Some *Fernandezia* flowers are tubular and brightly colored, suggesting pollination by hummingbirds. Others species (formerly *Pachyphyllum*) are white or translucent green. Once hummingbirds begin visiting slightly more colorful flowers in a variable population, those blossoms become reproductively isolated: they no longer swap genes with their drab neighbors, and evolution proceeds along separate tracks. The bright flowers strongly attract hummingbirds; the pale flowers appeal to insects. Those two groups eventually may become distinct species, despite being neighbors, a process called *sympatric speciation*. Genetic isolation more commonly is enforced by geographic barriers such as mountains or oceans (*allopatric speciation*), but in this case by the disparate preferences of pollinators.

Fernandezia ionanthera and column's sheathing wings
Franco Pupulin

F. subbiflora showing fleshy leaves and stem
Franco Pupulin

Galeandra

Description: A trumpet-shaped lip, floral spur, and cylindrical pseudobulbs distinguish *Galeandra*. **Flower:** Narrow, greenish or cinnamon sepals and petals are splayed upward above the lip. Rolled into a funnel, the lip presents a forward edge that curls down, while the rear portion forms a large, backward-pointing spur. The lip is white, with purple rays or a broad wash of magenta, and conceals a short column with a beaked tip. **Plant:** Tall, cigar-shaped pseudobulbs (to 12" [30 cm] tall or more) retain clasping bases of leaves that have dropped off. The few terrestrial species lack pseudobulbs. Arching leaves are thin and narrow, with visible veins, especially on their underside. The curving inflorescence from the pseudobulb apex yields up to a dozen flowers. **Similar:** See *Cyrtopodium*.

Distribution and Diversity: Nearly 40 species grow as low-elevation epiphytes, or rarely as higher-elevation terrestrials. They can be found from Florida and Mexico to Bolivia and Argentina; peak diversity lies in the Amazon rainforest.

Ecology and History: Euglossine bees have been found carrying *G. devoniana* pollinia, and the spurred flowers of the whole genus are likely pollinated by these mostly solitary bees. Some species inhabit forests with a pronounced dry season, dropping their leaves to reduce water loss and relying on stored nutrients in the columnar pseudobulbs. The aforementioned species was named to commemorate a giant in the orchid world, William Spencer Cavendish, Duke of Devonshire (1790–1858). Introverted because of his near deafness (and thus nicknamed the Bachelor Duke), he was spellbound by orchids and set about constructing the most lavish and sophisticated greenhouses in England. Previously orchids had been cultivated in hot, steamy brick houses heated by coal fires (see *Batemannia* and *Paphinia*). Innumerable orchids perished in these stultifying rooms until the Duke began growing his stunning collections in well-ventilated glass greenhouses that maintained distinct climates matching the needs of each orchid. The most renowned was the colossal Crystal Palace, more than 1800 feet long, built for the Great Exhibition of 1851 by Cavendish's chief gardener John Paxton. In addition to the many orchids named for him, the world's most popular banana variety is called the Cavendish.

Galeandra greenwoodiana
Andy Phillips

G. minax showing nectar spur
Andy Phillips

G. lacustris "cigar" pseudobulbs
Ron Kaufmann

Galeottia

Description: The tiger-striped or maroon flowers of *Galeottia* exhibit strong fragrances and toothed to deeply fringed lips. **Flower:** Large (to 4" [10 cm]), long-lived, fleshy blossoms have intricate, intoxicating perfumes. Petals and sepals are similar, greenish yellow with strong reddish-brown stripes, or wholly wine colored. The folded bases of the lateral sepals unite to form a chin. Deeply frilled side lobes of the lip flank a raised series of parallel, knife-like grooves; the front lobe is broad, usually fringed, and curled down at the tip. The underside of the arching, winged column is striped with purple. **Plant:** Clumped pseudobulbs each bear 2 or 3 broad leaves that are distinctly veined and weakly pleated. Inflorescences of 1 or few flowers arise from the leaf bases. **Similar:** See *Batemannia*. *Zygopetalum* sepals do not form a chin, lip side lobes are not frilly.

Distribution and Diversity: Approximately a dozen terrestrial and epiphytic species decorate low- to middle-elevation forests. The genus ranges from southern Mexico to Bolivia and Brazil.

Ecology and History: Although evidence is scarce, the shape of the lip, the height of the gap between lip and column, and the strong, complex fragrances all point to pollination by male euglossine bees (*Eulaema*). Males are attracted to a unique blend of scent compounds (see *Gongora*) they collect for use in reproductive displays. Analysis of *Galeottia* flowers identified at least 13 different fragrance components, ranging from rose to pine resin, geranium to wintergreen, and lilac to lemon. The genus is named in honor of Belgian scientist Henri Galeotti (1814–1858), born to French and Italian parents. Galeotti spent five years exploring the botany and geography of Mexico, where he and Jean Linden (see *Paphinia*) became the first to climb the flanks of Pico Orizaba. At 18,000 feet (5486 m), Orizaba is the third highest mountain north of Colombia; only Mount Logan in Canada and Mount McKinley in Alaska are taller. During his industrious visit he collected some 8000 plant specimens, many previously undescribed. About his herbarium he gushed, "I have gathered already a great number of vegetables . . . [an] infinity of beautiful and curious plants" (Ossenbach 2009: 78).

Galeottia grandiflora
Andy Phillips

G. negrensis
Andy Phillips

Gomesa

Description: *Gomesa*, the Little Man Orchids, are readily identified by fused lateral sepals that often resemble pants: the small flowers of some species evoke a person performing jumping jacks. **Flower:** Blossoms exhibit a wide diversity of forms, but nearly all feature the characteristic fused sepals. Petals mostly are raised, and the dorsal sepal is erect. The lip may be large and prominent, and usually is 3-lobed. Some columns are upright with a pale cap, mimicking the little man's neck and head. Flowers of several species are fragrant, sometimes powerfully so, with a citrus-like scent. **Plant:** Laterally flattened pseudobulbs (to 4" [10 cm] tall), occasionally cone-shaped, bear 2 leaves at the tip and often several leaf-like sheaths at the base. Leaves are soft, arching, and long (to 12" [30 cm]). Up to 30 flowers arise from a single short, curved inflorescence. **Similar:** *Rodriguezia* flowers have a nectar spur. *Oncidium* blossoms are bright yellow or reddish brown, and lateral sepals are unconnected.

Distribution and Diversity: Some 120 species now belong to *Gomesa*. They are primarily epiphytic, with some rock dwellers, and occur in middle-elevation humid forest in central Brazil (the core of their range), Paraguay, Uruguay, northeast Argentina, and Amazonian Peru.

Ecology and History: Some *Gomesa* flowers produce lipid-rich oils as a pollinator reward. Secreted onto the lip by glands near its base, the abundant oil attracts female *Centris* bees. These large-bodied bees are adapted to collecting oils from diverse plants, including non-orchids in the largely tropical Malpighiaceae family; *Gomesa* flowers often resemble malpig blossoms. Specialized combs on the bees' front 4 legs glean the oil, then transfer it to brushy patches on the hind legs. Females use oils to construct their nests, by sticking together bits of sand and wood, and to provide food for the larvae that emerge. These bees also pollinate Brazil nut trees (see *Coryanthes*) and cashew trees. Some *Gomesa* flowers exhibit an extreme adaptation to avoid self-pollination: the pollinia and the column of the same plant can literally destroy each other on contact. Studies of *G. bifolia* revealed that only 30% of self-pollinated flowers set fruit, compared with 100% of cross-pollinated flowers.

Gomesa recurva
Joe Meisel

G. planifolia inflorescence of "little man" flowers
Andy Phillips

Gomesa imperatoris-maximiliani (formerly *Oncidium crispum*)
Andy Phillips

G. praetexta (formerly *Oncidium praetextum*) in habitat
Ron Kaufmann

G. cornigera (formerly *Baptistonia cornigera*)
Ron Kaufmann

G. silvana (formerly *Baptistonia silvana*)
Ron Kaufmann

Gomphichis

Description: Flowers of *Gomphichis*, clustered on a spike, are distinct in their lip-uppermost orientation and dense coat of hairs. **Flower:** Small blossoms (1/2" [1.3 cm]) are pale green, yellow, and white. Narrow sepals are covered in woolly hairs, while petals sprout tiny, stiff protuberances: *Gomphichis* derives from the Greek words for "pin," or "nail," referring to the often rigid floral hairs. The lip, also hairy, is held uppermost (non-resupinate) over an S-shaped column and is commonly cleft in the center. **Plant:** Fleshy leaves are clustered at the soil surface, often in a rosette. Leaves narrow at their base but are not stalked. An erect inflorescence, packed with flowers and small leafy bracts, arises from the center of the leaves. **Similar:** *Prescottia* petals are smooth, lip is deeply pouched. *Pelexia* lip is presented below the column.

Distribution and Diversity: Some 2 dozen, mostly terrestrial species punctuate middle- to upper-altitude cool forests and alpine habitats. The genus, most abundant in the high Andes, ranges from Costa Rica to Venezuela and Peru.

Ecology and History: The small, pale flowers of *Gomphichis* are presumed to be pollinated by diminutive sweat bees (see *Cranichis*). Commonly growing in chilly, high-elevation sites, these plants possess a number of adaptations to survive the frosty climate of the mountains. First, like many terrestrial plants, their growing buds reside near or beneath the soil surface; the inflorescence and low-lying leaves emerge from these buds. Air temperature drops steeply even a few inches above the ground, but the sensitive buds will not freeze if they are insulated by the sun-warmed soil. Second, the flowers and inflorescence are covered in dense hairs, called *pubescence*. This woolly coat traps an insulating layer of air near the flower. As icy mountain winds roar past the plant, the downy flowers retain some of the day's heat which otherwise would be stripped away by the breeze. In experiments with other high-elevation plants (*Espeletia*), a layer of pubescence just one-twelfth of an inch thick kept plant tissues 10 to 30 degrees Fahrenheit (5–10 °C) warmer than the surrounding air.

Gomphichis inflorescence in high-elevation habitat
Ron Kaufmann

Gomphichis, close-up of flowers
Ron Kaufmann

Gongora

Description: The unusual upside-down flowers of *Gongora* resemble dragons in flight, with petals partly fused to the column. **Flower:** Lateral sepals flare and curl back but are cupped in some species; the dorsal sepal points down and is partly joined to the column. Tiny petals, like spatulas with pointed tips, are fused at their bases to the column sides. The convoluted lip is fleshy, waxy, somewhat earlobe-like, with finger or hair-like protuberances. Flowers are medium sized (2–3" [5–7.5 cm]), colorful, and mostly fragrant. **Plant:** Egg-shaped pseudobulbs are deeply grooved. A few long (to 12" [30 cm]), broad, and strongly pleated leaves spring from the pseudobulb tip. Several thick inflorescences hang from the pseudobulb, each flower's stalk also curving distinctively downward. **Similar:** *Stanhopea* flowers are much larger, with free petals (not attached to column); their pseudobulbs are squat and hardly furrowed. See also *Acineta*.

Distribution and Diversity: Approximately 70 species dwell as epiphytes in low- to middle-elevation wet forest, from Mexico to Bolivia and Brazil.

Ecology and History: Some *Gongora* plants grow in nests of ants, for example, the formidable trap jaw ant (*Odontomachus*), which aggressively defend the orchid. Other species provide hollow pseudobulbs perfect for ant colonies. Upward-pointing trash roots catch falling organic debris, providing nutrients typically scarce for epiphytes. Male euglossine bees pollinate the flowers, lured by an intricately blended scent. Each plant releases a unique combination of fragrance components, attracting just 1 of the more than 50 species of euglossines in the area, ensuring its pollen will be delivered only to flowers of the same species. As blossoms are notoriously similar, botanists often rely on "pollinator's judgment" to separate *Gongora* species. The bees collect scent from the lip's base, hanging precariously upside down. When departing compact flowers like *G. armeniaca*, they crawl backward down the column, contacting the pollinia. In open flowers like *G. quinquenervis*, the bee drops into flight and executes a simultaneous barrel roll, but often collides with the column, skidding onto the reproductive area. Flowers avoid self-pollination through sequential male and female phases, and encourage long-distance gene flow by synchronizing phases with other plants in the region.

Gongora ilense
Andy Phillips

Gongora convoluted lip (right), column (center) with fused petals
Ron Kaufmann

G. armeniaca
hanging inflorescence
Andy Phillips

G. chocoensis
Andy Phillips

Govenia

Description: Among terrestrial orchids, *Govenia* species are differentiated by leaves with long petioles and the concave petals and dorsal sepal that form a broad hood. **Flower:** Blooms vary in color from white to purple to tangerine orange, with bars and stripes common on the petals. Sickle-shaped lateral sepals project forward or bend slightly down. The short, fleshy lip is white or yellow, and spotted. The column arches upward, in tandem with the curled lip. **Plant:** One to 3 thin, pleated leaves arise from an underground storage organ (corm), commonly as an opposing pair. A distinct joint marks the union of leaf blade and elongated petiole, the latter obscured by coarse sheaths; often the base of one leaf encircles the other. Leaves are deciduous during dry months. An erect, spike-like inflorescence emerges from the sheath, bearing many small flowers.

Distribution and Diversity: More than 20 species occur on humid forest floors, from sea level to middle elevation. *Govenia* plants prefer rich, organic soil with rotting leaf litter. They range from southern Florida and the Caribbean to Bolivia and Argentina.

Ecology and History: The flower color, open form, purplish markings, and pleasant fragrance convincingly suggest pollination by bumblebees. The stripes serve as nectar guides, visually leading bees into the flower. Crawling between the lip and the column, they accomplish pollination. *Govenia* species are known to produce flowers surprisingly early in life, and the explanation may lie in their terrestrial lifestyle. Terrestrial orchids, unlike epiphytes, enjoy ready access to soil nutrients and consistent water. Their leaves are delicate, lacking the waxy surface used by other orchids to lock moisture inside their tissues. *Govenia* plants begin life as underground, coral-shaped *saprophytes*, meaning they consume decaying organic matter rather than produce energy by photosynthesis. Such saprophytes often rely on symbiotic fungi to perform the actual digestion of organic material, and in fact may be parasites on the fungus. Once sufficient nutrients have been amassed, the plant develops a storage corm and sends up its first leaves. In *Govenia*, this saprophytic phase allows the plant to accumulate energy rapidly, fueling the early production of flowers.

Govenia cf. *praecox* showing
hood of petals, dorsal sepal
Andy Phillips

G. liliacea
Andy Phillips

G. aff. *sodiroi* showing curved
column and lip
Andy Phillips

Guarianthe

Description: Showy purple or bright orange flowers with a tubular lip, and tall, club-shaped stems characterize *Guarianthe*. **Flower:** Large sepals and petals typically are spreading, the petals more broad and ruffled; flowers are 3 to 5" (7.5–12.7 cm] across. Color commonly is purple, rarely pale pink or white, with one orange species (*G. aurantiaca*), but is quite variable within a population or even on an individual plant. The lip is trumpet-shaped, its whitish base wrapping the small, blunt column. **Plant:** Tall pseudobulbs (to 18" [46 cm] or more) are cylindrical, but narrower at the base, especially in older plants; 2 or 3 leaves emerge from the tip. **Similar:** See *Cattleya*. *Rhyncholaelia* has only 1 flower per inflorescence.

Distribution and Diversity: Four species and 1 natural hybrid grow as epiphytes in lower- to middle-altitude wet forests. *Guarianthe* can be seen from Mexico to Venezuela, and in Trinidad.

Ecology and History: In 1939, Costa Rica chose as its national flower a stunning orchid known locally as Guaria Morada, "purple orchid." Originally named *Cattleya skinneri*, the species honors George Skinner, indefatigable explorer of Guatemala (see *Lycaste*). More than 60 years later it was assigned to a new genus, dubbed *Guarianthe* in recognition of the Costa Rican word for orchid, "guaria." While the species is widely grown in gardens and even on rooftops, it has been heavily overcollected and despite its national status is rarely encountered in nature. In nearby Mexico, studies of another favored guaria, *G. aurantiaca*, have shown that overcollection threatens the species with extinction. Flowering plants are stripped from their natural habitat, already under assault from agriculture, and sold by the basketful as ornamentals. Population viability analysis, a technique to estimate the resilience of plants or animals in the face of threats like hunting and collecting, has shown the population cannot sustain annual harvests of even 5% of wild plants. Critically, the practice of snatching flowering plants eliminates reproductive individuals from the population, crippling seed production and accelerating extinction. Sale of commercially propagated plants, rather than those stripped from forests, is the only responsible approach to ensuring these flowers continue brightening people's homes while reducing pressure on wild populations.

Guarianthe aurantiaca with short column (compare *Cattleya*)
Andy Phillips

G. skinneri, named for explorer George Skinner
Ron Kaufmann

Habenaria

Description: *Habenaria* flowers are unique in having a concave hood over the column and an elaborately lobed lip with a descending spur. **Flower:** Typically pale green or white, some species have yellow, orange, or lilac flowers. The dorsal sepal forms a rowboat-shaped hood, and often is joined to the bases of the narrow petals. Lateral sepals are slightly larger but also usually flattened and very narrow; the genus name derives from the Latin for "reins," after these strap-like features. The deeply lobed lip sometimes projects a spectacularly frilly margin, while the base forms a shallow or elongated nectar spur. The column is short and complex, with paired horns at the tip. **Plant:** Large underground tubers produce tall, unbranched stems (to 30" [76 cm] or more) bearing leaves that alternate up their length or lie flat around the base. Leaves are thin and delicate, variable in color, and deciduous.

Distribution and Diversity: Commonly called bog orchids, more than 800 species inhabit tropical and temperate grasslands, wetlands, and occasionally shady forest. They are distributed globally but are most diverse in Brazil and Africa.

Ecology and History: *Habenaria* floral spurs produce nectar to attract pollinators. Green and white flowers are visited by moths, while brightly colored species rely on hummingbirds. Thanks to the worldwide distribution and terrestrial habit of the genus, it has been widely encountered by humans, and innumerable uses have been recorded. In Malawi and other African nations, the tubers are mashed to form a meatless sausage called *chinaka* (aka chikanda), an important food for nearly all rural households. Conversely, the Shona people of Zimbabwe believe the powdered tuber, when eaten, curses enemies to a slow wasting death. This belief stems from the tendency of these annual herbs to senesce during dry seasons. In North America, the Iroquois and Ojibwe used *Habenaria* as a treatment for rheumatism, tuberculosis, infection, and even as an aphrodisiac. Botanists discovered that the waterspider bog orchid (*H. repens*) manufactures a protective compound (habenariol) to deter herbivory by crayfish. Terrestrial plants, including aquatics, face greater risks of herbivory than epiphtyes, and their widespread production of defensive chemicals may explain the abundant medicinal uses found for *Habenaria*.

Habenaria monorrhiza, nectar spurs, strap-like sepals
Ron Kaufmann

Habenaria in habitat
Ron Kaufmann

Close-up of hooded flowers
Ron Kaufmann

Houlletia

Description: *Houlletia* plants have compact pseudobulbs, pleated leaves, and showy flowers with delicately intricate lips. **Flower:** Two flower forms exist. One is spreading (e.g., *H. tigrina, H. odoratissima*), with the lip lowermost. Sepals and petals are yellow to orange with dense purple patterns. The lip's triangular lower region resembles a steer's head, while the upper part bears two backward-pointing, flipper-like side lobes. The second form (e.g., *H. sanderi and H. wallisii*) is bell-shaped, cream colored with fewer spots, and the column lowermost (non-resupinate). Their lips lack the triangular lower portion, and upper flippers are more enlarged and curling. All flowers are fragrant. **Plant:** Egg-shaped, clustered pseudobulbs produce a single leaf from the tip, and an arching or hanging inflorescence from below. Broad leaves are strongly pleated, with a distinct stalk. **Similar:** *Paphinia* pseudobulbs have 2 or more leaves, and flowers with a frilly lip. *Koellensteinia* pseudobulbs are tiny and hidden by leaf bases.

Distribution and Diversity: Some 10 species, terrestrials and a few epiphytes, decorate wet middle-elevation forest from Guatemala to Brazil and Bolivia.

Ecology and History: Male euglossine bees are attracted to the complex fragrances that include notes of wintergreen, cinnamon, eucalyptus, and vanilla. *Houlletia tigrina* is pollinated by *Eulaema meriana* bees, whose abdominal stripes resemble bumblebees (see *Aspasia*). Strong fliers, these robust insects may travel more than 30 miles (50 km) to receptive flowers. Such bees are uncommon above 6550 feet (2000 m), however, restricting the range of orchids that rely on them. *Houlletia sanderi* commemorates the world's greatest orchid grower, Henry Sander (1847–1920). His partnership with prolific collector Benedict Roezl (see *Miltoniopsis*) thrust his nursery business in St. Albans (England) to the zenith of the orchid world. Even St. Albans Cathedral is now adorned with carved orchids. Sander marshaled two dozen collectors to scour the globe and stock 60 massive greenhouses. He sold literally millions of orchids to royalty and commoners alike. Sander was uniquely responsible for the booming popularity of orchids, his enthusiasm for them so great a friend said, "He could turn a blacksmith into an orchid grower" (Reinikka 1995: 263).

Houlletia odoratissima, spreading form
Andy Phillips

H. sanderi, bell-shaped form, non-resupinate flower
Ron Parsons

Houlletia wallisii, bell-shaped form
Ron Parsons

H. tigrina showing lip side lobes
Andy Phillips

H. brocklehurstiana vegetation in habitat
Ron Kaufmann

Huntleya

Description: Stunning, star-shaped flowers with a checkerboard-like pattern of yellows and browns and a strongly fringed lip base are characteristic of the popular *Huntleya*. **Flower:** Large, open flowers vary in color, usually yellow or white near the center, giving way to strong purplish-brown markings farther out. Waxy sepals and petals are similar, pointed and arranged as a 5-parted star. The severely narrowed lip base bears a series of parallel ridges or keels, with pointed or bristly edges. The short column's tip is covered by a drooping hood. **Plant:** Narrow, flat leaves (to 12" [30 cm] long) evince distinct veins; a unique fold line crosses the leaf below the midpoint. Pseudobulbs are absent, and leaves emerge as a fan from a thick runner (rhizome). Inflorescences arise from the leaf bases, each bearing just 1 flower. **Similar:** See *Cochleanthes*. *Pescatoria* lip keels are rounded in front. *Stenia* lip base is pinched, cupped. *Warczewiczella* flowers are not star-shaped, keels lack bristles.

Distribution and Diversity: Approximately 15 species grow epiphytically in cool, damp habitats. They can be enjoyed from Costa Rica and Trinidad to Peru and Brazil.

Ecology and History: The penetratingly fragrant flowers of *Huntleya* attract euglossine bees. Male *Eulaema* bees crawl onto the fringed keels of *H. meleagris* and *H. burtii*, receiving pollinia on their heads from the overhanging column. *Huntleya* plants often develop aerial roots that hang beneath them, dangling in the breeze. Their thick, whitish outer covering, called *velamen*, is specially adapted to absorb moisture rapidly from infrequent rainfalls, soaking it up literally like a sponge. This feature is common to many epiphytic orchids, which lack a connection to moist soil. Instead, the velamen holds water that can keep the orchid hydrated until the next cloudburst. Orchid roots also typically assist with nutrient uptake from the surrounding environment: soil, tree bark, or rock faces. Nearly all orchids have symbiotic fungi in their roots that reach out into the substrate, helping to gather, process, and mobilize scarce nutrients (see *Govenia* and *Mormodes*). Studies of *H. burtii* revealed that these fungi are absent from the aerial roots, underscoring their function as collectors of water rather than nutrients.

Huntleya meleagris showing lip keels, overhanging column
Ron Kaufmann

H. fasciata showing leaf fold lines (beneath center flower)
Andy Phillips

Ionopsis

Description: *Ionopsis* means "violet-like" in Greek, a reference to the purple flowers with a broad lip (although only vaguely resembling violets). **Flower:** Even within a single cluster of plants, the small (to 1" [2.5] across), delicate flowers vary widely in color from deep purple to white. Sepals and petals are alike, although petals are less pointy; lateral sepals join at the base to form a shallow, sac-like depression. The much larger, rounded lip is flat or concave and deeply notched in some species. **Plant:** Tiny pseudobulbs, completely hidden by leafy sheaths, bear several leaves from the apex, or a single, scale-like leaf. Leaves are thickly leathery, with a crease above and pronounced keel below, or are cylindrical in cross section. The short column terminates in a 3-lobed cap. **Similar:** See *Comparettia*.

Distribution and Diversity: Just 6 epiphytic species are widespread from Florida and the Caribbean to Mexico and Paraguay, and even the Galapagos Islands. They favor low- to middle-elevation hardwood forests.

Ecology and History: *Ionopsis utricularioides* is a wide-ranging epiphyte on twigs of shrubs, forest trees, and fruit orchards. It matures rapidly, flowering in as little as 1 year, an adaptation to the unstable nature of its perch (see *Erycina* and *Rodriguezia*). Clustered plants can produce impressive floral displays but have low rates of fruit production. Hand-pollination experiments suggest one cause is limited pollinator visits, as artificially pollinated plants yield three times the fruits of unassisted plants. Nutrient limitation also may play a role: successfully pollinated plants invested considerable energy into fruit production but exhibited significantly reduced growth during the following year. Like nearly all orchids, *Ionopsis* plants rely on symbiotic fungi to support seed germination (see *Huntleya* and *Sobralia*). The particular fungal strains associated with *I. utricularioides* are unusually widespread throughout the American tropics, perhaps explaining why this orchid is found across such an enormous range. Although orchids traditionally are not considered parasites of their host trees, this particular species is known to strangle coffee bushes, from which it must be removed. Botanists suspect that fungal partners of other *Ionopsis* species actually invade their host trees, robbing them of nutrients.

Ionopsis utricularioides (pink color form) showing leaves
Andy Phillips

I. utricularioides (white color form)
Valérie Léonard

Brilliant *I. utricularioides* showing broad color variation
Ron Kaufmann

Isabelia

Description: *Isabelia* plants are easily recognized by exceptionally narrow leaves and fibrous sheaths on the pseudobulbs. **Flower:** Flowers are white, pink, or magenta in color, spreading open, with petals either broader or narrower than, but not similar to, the sepals; in *I. violacea*, however, the difference is minimal. The lip usually is larger and oblong, paler at the base in the pink and purple species. A shallow sac-like spur is formed where the sepals meet the base of the lip. The column is short and stubby. **Plant:** Clustered pseudobulbs are ovoid but obscured by a sheath of fibrous netting that brings to mind a tubular loofah sponge; the netting is replaced by papery sheaths in *I. violacea*. Each pseudobulb sports a single leaf, the entire plant growing in a compact, tangled scramble. The leaves are wiry, ranging from merely in-rolled to fully cylindrical, and quite short (2–3" [5–7.5 cm]). **Similar:** *Pseudolaelia* pseudobulbs typically are much larger and widely spaced (up to 10 cm apart), lack netting, and bear multiple leaves. *Laelia* is limited to Central America and the Caribbean.

Distribution and Diversity: Three species and 1 natural hybrid are known. All occur in lower- to middle-altitude forests of Brazil, and in northeast Argentina.

Ecology and History: This attractive genus is named after Isabel Augusta (1846–1921), daughter of Brazil's last emperor, Pedro II. Although Pedro despaired at having no surviving male heirs, Isabel was provided an excellent education and grew into a tough, intelligent, and forward-thinking woman. She also inherited from her father a lifelong passion for botany. Nicknamed "The Redemptress," Isabel was regent of the country when Brazil enacted laws ending slavery, becoming one of the world's last nations to do so. In 1888, she personally signed the Golden Law (Lei Áurea) freeing nearly a million Brazilian slaves. Upon hearing the news, overjoyed at the progressiveness displayed by daughter and country, Pedro shouted "Great people, great people!" (Calmon 1975). Unfortunately, some great people, wealthy merchants and plantation owners in particular, were apoplectic over the financial consequences of the new law, and they soon conspired to topple Brazil's last monarch.

Isabelia violacea
Andy Phillips

I. virginalis showing fibrous sheath
Ron Parsons

I. pulchella
Andy Phillips

Isochilus

Description: Petite tubular flowers and grass-like vegetation are diagnostic of *Isochilus*. **Flower:** Small flowers (to 1" [2.54 cm]) are pink, orange, or magenta. Sepals and petals are fused into a bell or tube, and bear small points on their 5 backward-curling tips. The lip closely resembles the sepals, albeit with paired deep purple spots near the tip. The genus name derives from the Greek words for "equal lip," referring to its similarity to the sepals. **Plant:** Lacking pseudobulbs, these plants grow in dense clumps of thin reed-like stems (to 2' [0.6 m] tall), drooping when overly tall. Narrow, flattened leaves arise on opposite sides of the stem, their bases overlapping as brownish sheaths. The short inflorescence emerges from the top, erect or arching, and bears a few densely clustered flowers also arranged opposite one another. **Similar:** *Jacquiniella* (not in this book) leaves are thick, sepal tips not pointed.

Distribution and Diversity: Just 12 species of epiphytes, terrestrials, and rock dwellers inhabit diverse habitats. The genus can be found from the Caribbean and Mexico to Brazil and Argentina.

Ecology and History: Unusual among orchids, the upper leaves often change colors to match those of the flowers (e.g., *I. major*). Some are pollinated by hummingbirds, as suggested by their bright colors and tubular shape (e.g., *I. aurantiacus*). Indeed, several species are called "sanguinarias" in Mexico for their blood-red coloration. *Isochilus linearis* is visited by the Rufous-tailed Hummingbird, a widespread denizen of tropical lowlands. This orchid was discovered in the West Indies and brought to England by none other than William Bligh, later gaining notoriety for his cruel treatment of the crew aboard HMS *Bounty*. His crew famously mutinied, set Captain Bligh adrift, and fled to Tahiti for refuge. The same plant has been shown to have antibacterial properties against *E. coli*, perhaps explaining its use in Mexico as a treatment for dysentery. Although some *Isochilus* species are pollinated by hummingbirds, many are capable of self-fertilizing. This process results in nearly all flowers successfully setting fruit, but the trade-off is that no new genetic material is passed to offspring, and adaptation to changing conditions is drastically slowed.

Isochilus linearis showing grass-like stems and leaves   *I. aurantiacus*   *I. major*
Andy Phillips   Andy Phillips   Andy Phillips

Kefersteinia

Description: *Kefersteinia* are distinguished by the presence of a keel-like tooth on the column and a fan-like leaf arrangement. **Flower:** Sepals and petals are similar, white to greenish yellow, frequently speckled with reddish purple. The concave lip usually is densely spotted, its outer edge often frilly and curling down, its base with a raised, fleshy hump on which the column rests. The robust column is shaped beneath like a boat hull, with a prominent keel. **Plant:** Small plants lack pseudobulbs, their narrow and lightly veined leaves borne on compact stems in a loose fan. Hanging inflorescences of a single flower arise from the leaf bases. **Similar:** See *Cochleanthes*. *Chondrorhycha* lateral sepals are swept back and tubular.

Distribution and Diversity: Some 70 species grow epiphytically, rarely terrestrially, at middle elevations in shady, wet forest from Mexico to Bolivia.

Ecology and History: Most *Kefersteinia* flowers are perfumed and pollinated by euglossine bees (usually *Euglossa*) in search of floral scents. The lip's central hump forces the bee to enter the flower from the side, and pollinia are affixed to the base of the bee's antennae; closely related *Huntleya* and *Cochleanthes* flowers avoid hybridizing with *Kefersteinia* by placing pollinia behind the bee's head. Studies of Mexican populations of *K. tinschertiana* suggest that a decline of pollinators in human-dominated landscapes hampers reproduction. Hand pollination of isolated plants might serve as a last-ditch conservation measure. That species' name commemorates German botanist Otto Tinschert (1915–2006), who spent years promoting the protection of Guatemalan orchids. Dismayed by the critical condition of that country's national flower, *Lycaste skinneri* (see *Lycaste*), he embarked on a vigorous greenhouse program to rear endangered species from seed. Although Tinschert strenuously lobbied diverse factions to create a national botanical garden, his dream remains unrealized. Another species, however, commemorates a botanical garden success story: *K. retanae* is named for Dora Emilia Mora de Retana (1940–2001), director of Costa Rica's Lankester Botanical Gardens (see *Chondrorhyncha*). During her tempestuous reign, she was instrumental in elevating Lankester to its current world-class status.

Kefersteinia taurina
Andy Phillips

K. gemma showing column tooth
Andy Phillips

Kefersteinia

Kefersteinia stevensonii showing faintly veined leaves and keeled column
Andy Phillips

K. tolimensis showing large callus that forces pollinators to approach column from side
Andy Phillips

Koellensteinia

Koellensteinia eburnea; see next page
Andy Phillips

K. eburnea leaves
Ron Kaufmann

Koellensteinia

Description: *Koellensteinia* plants are characterized by narrow, veined leaves and small, pale flowers with violet markings. **Flower:** Attractive white or muted yellow flowers (to 1" [2.5 cm]) are pleasantly fragrant. Similar petals and sepals, often marked with thin violet bars, project forward slightly. A fleshy, notched or toothed hump (callus) adorns the lip's base. Side lobes of the lip flank the callus, while the fore lobe is broadly spade-shaped. The short column, its underside also violet, usually has wings that meet the lip's side lobes, forming a short, mouth-like tunnel. **Plant:** Tufts of leaves arise from short stems that in some species thicken into pseudobulbs as they age, albeit obscured by sheaths. Each stem bears 1 to 3 leaves, grass-like or broader, pleated and stalked. An upright inflorescence with few to numerous flowers emerges from the leaf base. **Similar:** *Kefersteinia* leaves typically are broader and arranged in a fan; petals and sepals are elongated, lacking violet bars.

Distribution and Diversity: About 15 species are found in low- to middle-altitude wet forests, as terrestrials or less commonly epiphytes. They occur from Belize to Bolivia and Brazil.

Ecology and History: Euglossine bees are attracted to *Koellensteinia* flowers by their perfume. Crawling along the lip, a bee clambers over the callus and through the passageway between the wings of the lip and column; backing out, the bee scrapes the overhanging column. This mechanically orchestrated sequence ensures that the bee picks up pollinia from one flower and presses them into the stigmatic region of the next. The flower's violet markings are an adaptation to attracting pollinators in the understory: sunlight reaching the forest floor has been stripped of its red and blue wavelengths by photosynthesizing plants overhead, leaving violet among the few visible colors. Even epiphytic species favor sporadic, densely shaded canopy perches. A survey of a single nutmeg tree (*Virola michelii*) in French Guiana found 72 epiphytic species, half of them orchids, including a lone *K. graminea*. Nearly all the epiphytes were represented by very few individuals, while only four were abundant. This distribution is a classic example of how, in the tropics, rare is common and common is rare.

Koellensteinia ionoptera showing flanking lip wings; and see previous page
Andy Phillips

K. graminea showing notched callus
Andy Phillips

Laelia

Description: Stunning pink, purple, or white flowers with a paw-like column adorn *Laelia*. **Flower:** Sepals of the huge (to 6" [15 cm]) flowers are mostly narrower than the similarly colored petals, sometimes strongly twisted. The lip is 3-lobed, with lateral wings flanking or encircling the column. The lip base is marked with stripes of yellow or maroon. Long and narrow, the column terminates in 3 blunt lobes, like a cartoonish paw. **Plant:** One or 2 leathery, strap-shaped leaves with a central fold arise from each pseudobulb. A papery bract sheathes the base of the pseudobulb, while the unbranched inflorescence, frequently long and bearing 1 to many flowers, emerges from its apex. **Similar:** See *Brassavola*, *Cattleya*, and *Isabelia*. *Pseudolaelia* occurs only in Brazil.

Distribution and Diversity: Taxonomically complex, *Laelia* presently comprises some 2 dozen species from Mexico to Peru and Brazil, and in the Caribbean. They are on display epiphytically and terrestrially in humid and dry forests across a range of altitudes.

Ecology and History: Hard evidence of any *Laelia* pollinators is lacking, which is highly unusual for such a popular genus. Judging from the flower shape, coloration, and presence of nectar, however, large bees and hummingbirds are suspected. Reproduction of *L. rubescens* has been studied in Costa Rican forest patches using genetic techniques. Larger floral displays led to greater dispersal distances for pollinia, presumably by attracting pollinators from farther away. Nonetheless, big bunches of plants showed numerous examples of pollination by their nearest neighbors. The seeming paradox points to hummingbirds, which can cover miles in search of abundant food sources, but habitually visit all flowers in a bunch before moving on. In Mexico, patches of *L. speciosa* and *L. autumnalis* are visited by humans, who collect their pseudobulbs for preparation of traditional figurines to celebrate the Day of the Dead. Called *alfeñiques*, these candies are shaped into horses, cows, and fruits, as well as morbid skulls, skeletons, and coffins. Pseudobulbs are ground into a paste, mixed with sugar and eggs, and aged into a dough that can be molded into myriad shapes and painted. The festivities, held on November first and second, have their origin in Aztec rituals celebrating departed souls.

Laelia superbiens
Andy Phillips

L. autumnalis showing paw-shaped column
Andy Phillips

Laelia splendida showing lateral wings of lip
Ron Kaufmann

L. albida
Andy Phillips

L. rubescens
Andy Phillips

L. colombiana
Ron Kaufmann

L. speciosa showing pseudobulbs with papery
sheaths on upper bulbs
Andy Phillips

L. anceps, uncommon white color form, flower
typically lavender
Andy Phillips

Lepanthes

Description: Stems clothed in distinctive upturned, trumpet-shaped sheaths and tiny flowers (to 1" [2.5 cm] across) with colorful, elaborate petals make *Lepanthes* easy to identify. **Flower:** Colors vary widely, petals usually more flamboyant than sepals. Lateral sepals sometimes are partly fused. Smaller, 2- or 3-lobed petals are wider than they are long, flanking the column like a pair of ears. The lip often bears 2 lobes, wrapping or framing the column in front of the petals. **Plant:** Thin stems, each with a series of ribbed sheaths, lack pseudobulbs. Stiff leaves, sometimes attractively patterned or textured (e.g., *L. calodictyon*), arise from the stem apex, as does the inflorescence. Flowers mostly open one at a time, above the leaf or, more commonly, beneath it, leaving a distinctive fishbone-like pattern of stubs on the inflorescence. **Similar:** *Platystele* and *Pleurothallis* stems lack trumpets. *Stelis* inflorescence has several to many flowers open at once, lacks fishbone pattern. *Trichosalpinx* petals are longer than wide, unlobed.

Distribution and Diversity: More than 1000 species in this gigantic genus grow as epiphytes or terrestrials in a wide range of elevations and habitat types. Found from the Caribbean and Mexico to Brazil and Bolivia, they reach peak diversity in cool mountain forests.

Ecology and History: *Lepanthes* flowers are predominantly fly-pollinated. Some species, including *L. glicensteinii*, employ a pseudocopulation strategy that deceives male fungus gnats into attempting to mate with the flower. As the aroused gnat turns to press his abdomen against the lip, pollinia are glued to his underside. Many *Lepanthes* species have similar morphology, and this strategy may be widespread in the genus. Other species attract flies with calcium oxalate crystals on the flower surface, where their reflective shine is a visual attractant (see *Stelis*). Strangely enough, calcium oxalate crystals occur widely in plants, almost exclusively as chemical protection against herbivores. Typical crystals resemble needles or barbed spears and are produced within plant tissues, as in the houseplant *Dieffenbachia*. Animals that eat these plants get a mouthful of tiny, excruciating slivers not unlike fiberglass. In *Lepanthes* and *Stelis*, however, the crystals take on a different form, encrusting the petals like salt rime, and are safely edible as pseudonectar by potential pollinators.

Lepanthes horrida showing lobed petals
Andy Phillips

L. menatoi and fishbone inflorescence
Joe Meisel

Lepanthes gargantua
Joe Meisel

Lepanthes stems with trumpet-
shaped ("lepanthiform") sheaths
Valérie Léonard

Lepanthes flower emerging above leaf
Ron Kaufmann

L. melpomene
Joe Meisel

L. calodictyon showing patterned foliage
Andy Phillips

L. escobariana
Andy Phillips

Lepanthes flower with broad, shallowly lobed petals
Ron Kaufmann

L. calimae with deeply divided petals
Ron Kaufmann

L. glicensteinii
Franco Pupulin

L. cf. *felis* flower with long, horn-like petals
Ron Kaufmann

L. (Neooreophilus) pilosella in habitat
Ron Kaufmann

L. yubarta visited by potential pollinator
Ron Kaufmann

Leptotes

Description: Miniature *Leptotes* orchids, found almost exclusively in southeast Brazil, display cylindrical leaves with a central groove and proportionately large, appealing flowers (to 2.5" [6.3 cm] across) of white and purple. **Flower:** Sepals and petals are similar, narrow and arching forward, typically pale pink or white. The broad, eye-catching lip has wing-like lateral lobes and an upturned or frilly outer edge. Often flanked by the lip's lobes, the column is short and broad. **Plant:** Small, tubular pseudobulbs are nearly indistinguishable from the thick, cylindrical leaves. Leaves are pointed, sometimes resembling small goat horns, and have a channel-like groove on the upper surface. A short inflorescence bearing 2 to 12 flowers arises from the leaf base. **Similar:** *Brassavola* plants and flowers are much larger, never purple, with an arrowhead-shaped lip.

Distribution and Diversity: Approximately 10 species of *Leptotes* are concentrated in southern Brazil and neighboring Paraguay and Argentina. They dwell as epiphytes and terrestrials in lower-elevation humid forest.

Ecology and History: This appealing genus was established when *L. bicolor* was encountered in the Organ Mountains near Rio de Janeiro. Local residents discovered that its egg-shaped seed capsules have a high concentration of vanillin (see *Vanilla*) and have long used the dried pods to flavor milk and ice cream. The area, named for its highly eroded peaks resembling giant pipe organs, was declared a national park in 1939 to protect a superb example of Brazil's Atlantic Forest. Famed for its incredible diversity, this habitat is home to more than 20,000 species of plants, at least 40% of them endemic (found nowhere else). Isolation for millennia from South America's other great rainforests is responsible for the evolution of so many unique species. Unfortunately, sugarcane and coffee plantations as well as explosive urban growth have decimated this coastal treasure. Today, less than 10% of the original habitat remains, and resident species, including orchids, are highly vulnerable. Support for conservation of the Atlantic Forest, however, has been galvanized around charismatic species such as the Golden Lion Tamarin monkey. Thanks to popular outcry and public support, state and federal reserves now protect some 9000 square miles (23,300 km²) of this incomparable forest.

Leptotes bicolor
Andy Phillips

L. harryphillipsii with grooved, cylindrical leaves
Andy Phillips

L. mogyensis showing tubular pseudobulbs
Andy Phillips

Liparis

Description: *Liparis* owes its name to the Greek word for "greasy," a reference to the shiny, oiled appearance of the leaf surface. **Flower:** Small flowers on an erect stalk are usually drab green, yellow or purple. Sepals and petals have in-rolled edges and a tubular appearance; the narrow petals can appear thread-like. Lateral sepals often are partly fused. The largest element is the lip, frequently notched or pointed at the tip and usually exhibiting wavy margins. The column is elongated and arching, with paired wings near the tip. Some species emit a putrid odor. **Plant:** Subterranean storage organs (corms) bear 1 to 3 thin leaves, most commonly 2. **Similar:** *Malaxis* flowers have a short, stubby column with the lip usually presented above it (non-resupinate).

Distribution and Diversity: More than 300 species occur globally, in a wide range of habitats and elevations, with the greatest diversity in tropical regions. Largely terrestrial, they include some rock-dwelling and epiphytic species.

Ecology and History: Flowers of *Liparis* are pollinated primarily by mosquitoes and flies, attracted by their unpleasant odor. Compared with bees, however, flies are unreliable pollinators. Whereas bees move efficiently from one flower to another collecting food (nectar or oil), flies move haphazardly, often visiting the same flower again and again. Perhaps because of this wayward behavior, many *Liparis* species are capable of self-fertilization. Studies of *L. loeselii*, a European and North American species, revealed that rainfall can induce self-pollination. Raindrops strike the column, loosening pollen masses that are washed onto the stigmatic area, fertilizing the plant. Because fly activity is low during heavy rains, this adaptation may improve reproductive success in particularly wet habitats. When seeds are produced, they must encounter a specific soil fungus (mycorrhiza) to germinate successfully; such is the case with most orchids. *Liparis* seeds, unlike those of most orchids, remain viable for up to 4 years, awaiting an encounter with the proper fungus. The extreme specificity of this relationship underscores the difficulty of reintroducing or transplanting orchids, especially terrestrial species: if the fungus does not co-occur, the plant will be incapable of reproduction.

Liparis nervosa
Franco Pupulin

Liparis with (uncommon) pleated leaves
Ron Kaufmann

Lockhartia

Description: *Lockhartia* plants are known as Braided Orchids, for the overlapping leaf arrangement that obscures the stem. **Flower:** Small flowers (<1" [2.5 cm]) are yellow or white with brownish-red markings. Lateral sepals often bend backward behind the lip, while similar petals curl forward. The complicated lip has paired lobes at its base partly surrounding the column; the forward part of the lip has a ruffled edge, or 2 to 4 lobes. Broadly winged, the column is short and stubby. **Plant:** Distinctly flattened stems are upright or hanging; pseudobulbs are absent. Small, leathery, and nearly triangular leaves are flattened laterally, with overlapping, clasping bases that cover the stem entirely. Short inflorescences emerge in small bunches from the upper leaf axils. **Similar:** See *Dichaea* and *Fernandezia*.

Distribution and Diversity: Approximately 30 epiphytic species prefer wet lowland forest. They can be found from Mexico and Trinidad to Bolivia and Brazil.

Ecology and History: Like several related genera, *Lockhartia* flowers manufacture oil that is collected by the bees serving as pollinators (see *Oncidium, Ornithocephalus,* and *Rudolfiella*). *Lockhartia micrantha* and *L. oerstedii* are pollinated by several species of *Centris* bees, which also obtain oils from non-orchid flowers (see *Gomesa*). Once pollinated, seed capsules form and eventually split along 6 seams to release microscopic seeds. Unlike other orchids, in which seams only partially open, *Lockhartia* capsules distinctively split all the way to their tip, perhaps to more fully disperse the contents. *Lockhartia pittieri* commemorates Swiss scientist Henri Francois Pittier (1857–1950), who comprehensively surveyed the flora of Costa Rica and helped establish its National Herbarium. He later convinced Venezuela to declare its first national park, now named in his honor. *Lockhartia chocoensis* is named after, and found in, an exceptionally biodiverse area of western Colombia called the Chocó. The region supports dense, lowland forests separated from the Amazon basin for millions of years by towering Andes mountains to the east. This isolation led to the emergence of an unusually large number of unique species (25% of plants, 30% of reptiles, and 45% of freshwater fish are endemic), which in turn has prompted declaration of the Chocó as an international Biodiversity Hotspot.

Lockhartia longifolia
Andy Phillips

L. longifolia foliage and
flowers
Ron Kaufmann

L. oerstedii inflorescence
Joe Meisel

Lycaste

Description: Large *Lycaste* plants combine broad, pleated leaves with showy, triangular flowers. **Flower:** Lateral sepals often are perpendicular to dorsal sepal. Petals are shorter and broader, bending forward but with tips commonly curling out. The lip has lateral lobes flanking the column, a broad central lobe bending down and back, and a curb-like undulation between the two regions. The thin column is long and slightly bent. Many species are fragrant, with pleasant cinnamon to peculiar soap-like scents. **Plant:** Densely clustered pseudobulbs are stout, egg-shaped, and often irregularly ridged. Several large (to 48" [1.2 m] long, 6" [15 cm] wide) leaves arise from the pseudobulb apex and base; leaves are thin, prominently pleated, and lack a stalk. Most yellow-flowered species are deciduous, retaining sharp defensive spines on the pseudobulbs. Wiry inflorescences, erect to horizontal, usually bear single flowers. **Similar:** See *Anguloa*. *Sudamerlycaste* flowers have frilly lips. *Xylobium* inflorescence bears multiple flowers.

Distribution and Diversity: Approximately 35 species of *Lycaste* adorn low- to high-altitude forests as epiphytes, terrestrials, or rock dwellers. They occur from the Caribbean and Mexico to Bolivia.

Ecology and History: *Lycaste* blooms are pollinated by male euglossine bees (most Central American species) and night-flying bees (Andean species), both seeking fragrances. In *L. aromatica*, the lip's lobes guide bees under the column to receive pollinia mounted on an erect stalk. After removal, the stalk bends forward as it dries, ensuring contact with the stigmatic area of the next flower. Upon pollination, the lip wilts rapidly, denying bees a landing platform. *Lycaste skinneri*, the national flower of Guatemala, was discovered by Scotsman George Ure Skinner (1804–1868), a tireless explorer who discovered nearly 100 new species. A contemporary remarked that Skinner "laboured almost incessantly to drag from their hiding-places the forest treasures of Guatemala. . . . there is scarcely a sacrifice which he has not made, or a danger or hardship he has not braved" (Reed 1956: 29). The pure white form, *L. skinneri* var. *alba*, became so sought after that it was collected virtually into extinction. Pollinator populations also plummeted because of United States Department of Agriculture insecticide spraying in Guatemala, which significantly reduced fruit set. Thus, although this breathtaking species is common in greenhouses worldwide, few wild populations survive.

Lycaste skinneri (variety *alba*)
Andy Phillips

Lycaste × *lucianiana* showing pleated leaf
Joe Meisel

Lycaste aromatica showing pseudobulbs with small spines
Andy Phillips

L. fuscina
Andy Phillips

L. deppei
Andy Phillips

L. consobrina
Andy Phillips

Lycaste showing leaves and pseudobulb
Ron Kaufmann

L. macrophylla showing lip lobes flanking column
Andy Phillips

Lycomormium

Description: The fanciful, some say grotesque, flowers of *Lycomormium* resemble *Peristeria* but differ in having an immobile lip. Plants can be very large, with stout pseudobulbs and long, pleated leaves (to 36" [91 cm/0.9 m] or more). **Flower:** Waxy, meaty flowers open only partly, forming spherical cups. Cream or ivory in color, all flower parts are densely speckled with wine or magenta. Petals are slightly smaller than sepals, both bluntly pointed. The pouched lip has upraised, keel-like side lobes and a crenulated frontal rim, resembling an upside-down football helmet. The lip is rigidly attached to the base of a short column. **Plant:** Pear-shaped pseudobulbs, their bases clasped by papery, tan sheaths, produce 2 to 3 leaves. Thick inflorescences bearing several flowers hang beneath the plant. **Similar:** *Peristeria* lip is hinged, not rigid.

Distribution and Diversity: About a half-dozen epiphytic and terrestrial species haunt Panama, Colombia, Ecuador, and Peru. They prefer wet middle-elevation forest.

Ecology and History: The strange, almost terrifying form of *Lycomormium* flowers is reflected in the genus name, drawn from Greek words meaning "like a wolf boogeyman": the pouched lip resembles a goblin's chin, the undulating margin a wolf's toothy snarl. The complex flower shape in reality is a highly evolved apparatus to accomplish pollination by bees. Analysis of two species (*L. fiskei* and *L. squalidum*) revealed the presence of diverse, fragrant chemicals (e.g., cineole, eugenol, limonene; see *Polycycnis*) known to attract male euglossine bees. The bees scrabble at the interior of the pouched lip to collect these scent compounds. When struggling to climb out of the smooth, waxy bowl, their escape route is meticulously guided by the lateral keels. As they clamber roughly over the lip's rim, they thrust their backs against the overhanging column, fertilizing the plant. This delicate ballet of sequenced movements to achieve pollination is repeated in other close relatives, *Stanhopea*, *Polycycnis*, and *Coryanthes*, all of which rely on male euglossines for reproduction.

Lycomormium fiskei in habitat
Ron Kaufmann

L. fiskei
Ron Kaufmann

Macroclinium

Description: The small twig orchids of *Macroclinium* are characterized by fans of leaves bearing delicate, spidery flowers with arrowhead-shaped lips. **Flower:** Tiny (to 1" [2.5 cm]) flowers are nearly translucent, pale green to white with a rich blush of lavender. Narrow, elongated sepals and petals are similar, spreading and drooping slightly. Shaped like a stalked arrow, the lip often is marked with purple splotches. The slender column bears an anther cap (enclosing the pollinia) on its upper surface rather than the tip, and frequently is bent back sharply. Blossoms are short-lived and lightly fragrant. **Plant:** The leathery leaves of these diminutive plants (to 4" [10 cm] across) often are warty with reddish speckling. Arranged in imperfect fans, leaves are flattened laterally (i.e., by squeezing the sides together). Pseudobulbs are small or absent. Typical inflorescences are branched, bearing several to many flowers. **Similar:** *Erycina* and *Ornithocephalus* leaves are smooth and green; *Ornithocephalus* roots are hairy. *Notylia* leaves are not arranged in fans.

Distribution and Diversity: Approximately 40 epiphytic species favor lower-elevation warm and humid forest. They range from southern Mexico to Peru and Brazil.

Ecology and History: Male euglossine bees visit the small flowers of *Macroclinium*, gathering fragrances. While grasping the flower, the bee's head bumps the column, and pollinia are glued between his eyes. When entering a subsequent flower, the forward-projecting pollinia are thrust into the column, and the flower is fertilized. *Macroclinium* plants commonly grow on twigs, in intact forest or on cultivated trees and shrubs, such as coffee, guava, and citrus. They limit water loss in hot climates by shutting their leaf pores (stomata) during the day, opening them only at night to inhale the carbon dioxide needed for photosynthesis. This technique is known as CAM photosynthesis, and although it occurs in only a tenth of all orchids, it is common to plants in hot, dry locations. One such dry region, the western Ecuadorian province of Manabí, is home to *M. manabinum*. The seasonally dry forests of this coastal zone are severely threatened by cattle ranching and agriculture, having lost 98% of their original extent. Dedicated local and international conservationists are fighting, however, to protect the remaining fragments of habitat.

Macroclinium manabinum, from western Ecuador
Ron Parsons

M. aurorae showing lip shape
Andy Phillips

M. aurorae inflorescences and fans of leathery leaves
Andy Phillips

Malaxis

Description: The small terrestrial plants of *Malaxis* can be recognized by the non-resupinate (lip-uppermost) flower orientation. **Flower:** Minute flowers are plain colored, typically green, occasionally orange, brown, or dull purple. Sepals curl back, as do the much narrower petals. The lip often is notched at the tip, while the column is short and stubby. **Plant:** Thin, soft leaves are visibly veined and typically sport undulating edges. One or more leaves are borne on a stem arising from an underground pseudobulb or onion-shaped corm. The inflorescence is erect, often terminating in an umbrella-like spray of flowers. **Similar:** See *Liparis*.

Distribution and Diversity: At present nearly 180 species of this taxonomically complicated genus occur. *Malaxis* plants are primarily terrestrial, with some epiphtyes, in a wide range of habitats on every continent (except Antarctica).

Ecology and History: Commonly called Adder's Mouth Orchids for the shape of the leaf, *Malaxis* plants typically grow in boggy conditions, and their flowers are pollinated by mosquitoes and small flies. The reflexed shape of the sepals is believed to permit these flies easier access to the flower's interior. Most *Malaxis* flowers do not exhibit the lip-below-column orientation typical of most orchids. The majority of orchid flowers commence life, in the bud, with the lip uppermost, above the column: during development, the flower twists 180 degrees so the lip is lowermost, a process called *resupination*. This orientation evolved to provide insect pollinators with a suitable landing platform, the often enlarged and runway-like lip. A handful of orchid groups are non-resupinate (e.g., *Paphinia*, *Scaphosepalum*, and complex flowers like *Gongora* and *Cycnoches*), having failed to complete this developmental twist. *Malaxis* flowers, however, are unique in achieving a non-resupinate orientation by rotating a full 360 degrees during maturation. The resulting topography requires pollinators to walk above, rather than below, the column and largely ensures that pollinia will be glued to the fly's belly. This uncommon orientation and pollinia placement help maintain the orchid's reproductive isolation from other, resupinate flowers that can be fertilized only by pollinia protruding from an insect's back.

Malaxis cf. *parthoni* in habitat
Ron Kaufmann

Malaxis showing non-resupinate flowers
Franco Pupulin

Masdevallia

Description: Three-parted and often tubular flowers with threadlike tails characterize *Masdevallia*. **Flower:** Ranging widely in size (1–15" [2.5–38 cm] between
outstretched tail tips), flowers are variable in color, typically reddish purple, white,
or orange, often with dark spotting or radial streaks. Showy sepals are partly
fused, forming a tube at the base, usually flaring broadly and terminating in slender filaments (tails). Petals are much smaller, often hidden within the tube, as are
the tiny lip and column. **Plant:** Diminutive plants lack pseudobulbs. Dense clusters
of leathery leaves emerge from short, branching stems. Fronds narrow to a stalk,
their base often sheathed by dry, leafy bracts. Inflorescences arise from leaf bases,
bearing 1 to 11 flowers. **Similar:** See *Dracula*.

Distribution and Diversity: More than 500 epiphytic and terrestrial species of
Masdevallia ornament low- to middle-elevation wet forests, often in shady locations. They occur from Mexico to Bolivia and Brazil.

Ecology and History: While a few *Masdevallia* flowers are hummingbird pollinated (e.g., *M. limax*), most are fertilized by flies and exhibit ingenious modifications to attract them. The long sepal tails catch the attention of insects in flight,
directing them to land on the broad center of the flower or to use the tails as lad-
ders to climb there. Spots or radiating stripes of purple guide the flies into the
floral tube, where they search for nectar, picking up pollinia in the process (e.g.,
M. hercules). Several species emit musty odors, often produced by glands in the
sepal tails: *M. nidifica* smells of decaying mushrooms, while *M. caudata* releases
a musk-like scent. These two entice flies that seek rotten material on which to lay
eggs. Like other fly-pollinated orchids, individual *Masdevallia* species have very
restricted ranges. The flies' limited flight capacity isolates populations, limiting
range expansion and promoting speciation. Up to 75% of all species are known
only from a single collecting locality. Such a distribution is a recipe for extinction
if the lone site is deforested. At least 2 species are thought to be extinct already due
to habitat loss: *M. lychniphora* in Peru and *M. menatoi* in Bolivia.

Masdevallia coccinea, example of
flat, open flowers
Andy Phillips

M. caesia in habit, showing hanging,
narrow-stalked leaves
Ron Kaufmann

M. infracta showing petals, column,
and lip
Andy Phillips

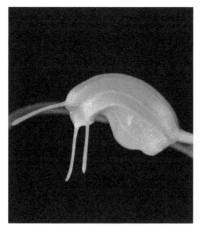

Masdevallia limax, hummingbird-pollinated
Andy Phillips

M. hercules, fly-pollinated
Ron Kaufmann

M. dynastes, example of flowers with recurved sepals
Andy Phillips

M. caudata, luring flies with rotten scent
Ron Parsons

M. ventricularia, example of flowers with bent tube
Joe Meisel

M. pumila in cloud forest habitat
Ron Kaufmann

Maxillaria

Description: *Maxillaria* owes its scientific name to the flower's distinctive lip, resembling a jawbone (maxilla); common names of Flame or Tiger Orchid refer to the frequently dazzling colors. **Flower:** Usually small, but occasionally large (to 7" [17.8 cm]) and showy, flowers are highly variable in shape and color. In most species, lateral sepals are arranged perpendicular to the similar dorsal sepal in a characteristic inverted T-shape. Smaller petals of the same color are held closer to the column. Side lobes of the oblong lip curl partially around the short, arching column. Many species exude fragrances ranging from sweet to putrid. **Plant:** Nearly all species have smooth, flattened pseudobulbs, sometimes well-spaced along cane-like stems; each pseudobulb bears a single leaf from the tip and is clasped by fresh green or dry brown sheaths at the base. Leaves typically are narrow, dark green, and leathery. Thick inflorescences with one flower and several leafy bracts emerge, in bunches or singly, from sheaths of new shoots. **Similar:** See *Bifrenaria*, *Cryptocentrum*, and *Trigonidium*.

Distribution and Diversity: *Maxillaria* comprises 200 to 300 species, depending on the classification system. The genus inhabits a wide range of elevations and ecosystems from Florida, Mexico, and the Caribbean to Brazil and Argentina.

Ecology and History: *Maxillaria* flowers are pollinated by stingless meliponines, larger euglossines, and other bees that scour blossoms for food. A few tubular flowers are visited by hummingbirds; others are wasp-pollinated. While no species offer nectar, many provide pseudopollen in the form of microscopic hairs on the lip (see *Polystachya*). Bees scrape off these powdery hairs, rich in protein and starch, and pick up pollinia during the scrabbling harvest. *Maxillaria* currently is undergoing significant taxonomic revision. DNA analysis (see *Trisetella*) revealed that the genus, as understood for more than a century, contains a large number of distantly related species. In taxonomic parlance, the former genus (called *Maxillaria sensu lato*, or "in the wider sense") is polyphyletic, encompassing both distant relatives and the core species (*Maxillaria sensu stricto*, "in the stricter sense"). Proposals to revive some old genera and to create new ones to accommodate the misfit species are gaining popularity among specialists. Such debates are beyond the scope of this book, however, and we largely restrict ourselves here to the core group.

Maxillaria lehmannii
Joe Meisel

M. arbuscula, compact flowers
Andy Phillips

M. weberbaueri showing lip shape
Andy Phillips

Maxillaria curtipes showing pseudobulb
Andy Phillips

M. fulgens, example of bell-shaped flowers
Andy Phillips

M. cf. *turkeliae*, with flowers up to 4"
(10 cm)
Ron Kaufmann

M. triloris, example of down-curling sepals
Ron Kaufmann

M. aurea, a climber, showing open
capsules
Ron Kaufmann

M. acuminata showing clumping growth
Ron Kaufmann

Maxillaria fractiflexa, example of spidery flowers
Ron Kaufmann

M. carinulata showing climbing rhizome, pseudobulb
Andy Phillips

M. scalariformis, another climber
Andy Phillips

M. cf. *splendens* flowers from pseudobulb bases
Andy Phillips

M. picta showing bracts below flowers
Ron Kaufmann

M. molitor, orange color morph
Ron Kaufmann

Microchilus

Description: Among terrestrial plants, *Microchilus* can be recognized by the spiraled inflorescence of downy flowers each bearing a nectar spur. **Flower:** Small, thin flowers have white to brownish-green sepals and petals, usually cloaked in short, pale hairs. The prominently white lip has 2 lobes, resembling an anchor or mustache, and a sac-like nectar spur at its base. Thin and short, the column is nearly hidden within the flower. **Plant:** Soft leaves range from green to brown and often are attractively variegated with silvery highlights. The hairy inflorescence arises from the leaf bases.

Distribution and Diversity: Nearly 100 species grow terrestrially or on rocks, occasionally as epiphytes. Plants prefer low- to middle-altitude forests, and occur from Mexico and the Caribbean to Brazil and Paraguay (and the Galápagos Islands). *Microchilus* recently was separated from *Erythrodes* to contain all species in the American tropics, while the latter retains the Old World species.

Ecology and History: The presence of a nectar-producing spur leads researchers to believe that bees and butterflies are the pollinators; however, hard evidence is lacking. Of greater interest is the discovery of the pollinator of a very ancient relative, the oldest definitive evidence of an orchid on Earth. A bee preserved in solid amber was unearthed in the Dominican Republic in 2000, with clearly recognizable pollinia glued to its back. This incredible fossil was dated to the Miocene epoch, some 15 to 20 million years ago. The ancient orchid, since gone extinct, is closely related to *Microchilus plantagineus*, a species found in the Caribbean to this day. Like the popular film *Jurassic Park*, the discovery of genetic material from a prehistoric orchid conjures outlandish possibilities. Perhaps one could create a Miocene Botanical Garden, reviving these primeval plants for orchid tourists to enjoy. Although dinosaurs had already disappeared from the planet, the theme park's menace could be provided by the gigantic alligator *Purussaurus*, more than 40 feet (12 m) in length, which lived in South America during the same period. Unfortunately, the modern species *M. plantagineus* is considered threatened by habitat loss throughout its range and soon may require non-imaginary parks for its protection.

Microchilus aff. *oroensis* close-up of hairy inflorescence
Andy Phillips

Microchilus in habitat, showing variegated foliage
Ron Kaufmann

Miltonia

Description: *Miltonia* flowers resemble closely related *Miltoniopsis* and *Oncidium* but have a distinct geographic range and simpler lip form. **Flower:** Fragrant, spreading, large (to 4" [10 cm] across) flowers are yellow, usually with dense brown barring, or white with a lavender or purple blush. Sepals and petals are alike, sometimes with pointed tips or ruffled margins. The showy lip ranges from fiddle-shaped to pansy-like, often bearing radiating stripes of purple or yellow. Its tip is often pointed, its base with several low, linear keels. Short and erect, the column has a blunt apex embellished by abbreviated wings. **Plant:** Flattened, oval pseudobulbs, flanked by leaf-like sheaths, bear 1 to 2 leaves from the top. Leaves are narrow, pale green, and folded near the base. Inflorescences with 1 to many flowers arise from the sheaths. **Similar:** See *Aspasia. Miltoniopsis* occurs in higher elevations of Central America and northwest South America. *Oncidium* lip projects arm-like side lobes.

Distribution and Diversity: Roughly 20 epiphytic or rock-dwelling species favor low-elevation montane forest. The genus occurs primarily in southern and eastern Brazil, with a few species spilling into Paraguay and Argentina.

Ecology and History: The perfumed flowers of *Miltonia* are bee-pollinated, but some uniquely so. The bright white lip of a few species is easily visible at night and attracts nocturnal bees that pollinate in total darkness. Taxonomists have labored to unravel a Gordian knot of species that resemble *Oncidium*, to which *Miltonia* orchids were originally assigned. Various morphological features, particularly the shape of the lip, along with molecular evidence have been employed in the struggle. As currently understood, *Miltonia* are essentially limited to Brazil, having been separated from the Andean and Central American species of *Miltoniopsis*. Making matters more complex, *Miltonia* flowers have been blended artificially with *Odontoglossum* (now *Oncidium*) and *Cochlioda* to form trigeneric hybrids known commercially as *Odontioda*. A quadruple genus hybrid added *Brassia* to the mix, and a quintuple hybrid tossed in *Oncidium* for good measure. Little wonder then that the reaction of taxonomist John Lindley (see *Barkeria*) to the first man-made orchid hybrid, bred by John Dominy in 1865, was to cry, "You will drive the botanists mad!" (Rolfe 1887: 166).

Miltonia cuneata showing lip keels
Andy Phillips

M. flavescens
Andy Phillips

Miltonia

Miltonia clowesii
Ron Kaufmann

M. moreliana showing flattened pseudobulb and leaves
Ron Kaufmann

Miltoniopsis

Miltoniopsis phalaenopsis; see next page
Ron Kaufmann

M. vexillaria (a bright color form), in habitat
Ron Kaufmann

Miltoniopsis

Description: *Miltoniopsis* orchids are characterized by spreading, white to purple, pansy-like flowers resembling some lower-elevation *Miltonia*. **Flower:** Showy blossoms are flat and wide, the broad sepals and petals white or lavender with red or pink markings. The large, spreading lip bears small horns at its base that curl up around the column. Often marked with red or purple radiating lines, the lip exhibits toothed or rib-like projections on its base. The column is short, with a distinct terminal cap. **Plant:** Tightly clustered pseudobulbs are flattened laterally (imagine a peach pit), each with a single leaf from its tip and additional leaves arising from the base. Leaves are soft and narrow, with papery sheaths retained at the bottom. The inflorescence arises from the pseudobulb base, bearing up to 10 flowers. **Similar:** See *Miltonia*.

Distribution and Diversity: Five species of *Miltoniopsis* reside epiphytically or terrestrially in the cool highlands of Costa Rica, Panama, and the Andes from Venezuela to Peru.

Ecology and History: *Miltoniopsis roezlii* commemorates an infamous orchid collector whose name is synonymous with the overharvest of orchids and wanton destruction of forests. Czechoslovakian Benedict Roezl (1823–1885) spent decades in Central and South America stripping orchids from the wild. He sent as many as 1 million plants to Europe, boasting of having plundered some 8 *tons* from Venezuela and Panama alone. His collection method was savagely simple: orchid-bearing trees were unceremoniously felled until whole forests had been obliterated. Sea voyage conditions being what they were, few of the delicate plants would survive the journey; it was said that one tree was lost for every "three scraps" of orchid material (Boyle 1893: 71). During a demonstration in Cuba of a fiber-spinning machine he had invented, the heaving gears seized his arm and destroyed his left hand. He bore a pirate-like hook thereafter, which scattered in terror more than one band of trail robbers seeking to ambush him. Roezl is commemorated today by a Prague statue that hoists an orchid with one hand and clasps a book with the other; evidently the sculptor chose to depict him before the mishap in Havana.

Miltoniopsis vexillaria; and see previous page
Andy Phillips

M. roezlii
Andy Phillips

Mormodes

Description: *Mormodes* are known as the Goblin Orchids or Flying Bird Orchids for their unusual twisted columns and lips. **Flower:** A male phase precedes the female phase, a sequence known as *gender diphasy*. Wide (to 6" [15 cm]), fragrant flowers are bizarrely shaped and heavily spotted. Petals and the dorsal sepal curl forward, hooding the lip and column. The saddle-shaped lip has large, downward-flaring side lobes. The slender column, in male phase, is twisted abruptly so the dorsal apex rests against the lip, exposing the pollinia. Flowers from which pollinia have discharged develop into the female phase, the column straightening. **Plant:** Tall (to 15" [38 cm]) spindle-shaped pseudobulbs appear segmented. Many broad, pleated leaves emerge from the seams (nodes) between segments and are deciduous. The inflorescence arises from the middle of the pseudobulb.

Distribution and Diversity: Some 80 species haunt wet lowland forest from Mexico to Bolivia and Brazil.

Ecology and History: The genus *Mormodes* is closely related to *Catasetum* and *Cycnoches*, but according to Calaway Dodson (see *Stenia*) its species are "in the process of developing unisexual flowers." Most species rely on gender diphasy to prevent self-pollination: the stigmatic zone is not receptive until after the pollinia are removed. Like their cousins, *Mormodes* flowers fire spring-loaded pollinia at male euglossine bees when the anther cap is touched; they also have floral fragrances that disappear upon pollinia removal or fertilization. All 3 genera manifest delayed readiness of their pollinia. When discharged, pollinia retain the protective anther cap, and their supporting stalk is tightly curled. Over 30 to 45 minutes the cap falls off and the stalk unbends, both actions permitting pollinia to be inserted into a receptive, female-phase column. These genera also share adaptations to a seasonally dry climate. They are deciduous and store starches and water in large, fleshy pseudobulbs. During the leafless period, many rely on symbiotic fungi to deliver nutrients to their roots from the substrate, often soil humus or decaying wood. Plants also have exposed, upward-pointing "trash roots" that catch falling leaves, twigs, and animal feces, which decompose and release precious nutrients.

Mormodes buccinator (male phase)
Ron Kaufmann

M. andicola showing pseudobulbs and foliage
Ron Kaufmann

Mormodes buccinator, yellow color form
Ron Kaufmann

M. hookeri showing twisted columns
Ron Kaufmann

M. maculata
Andy Phillips

M. wolteriana showing pleated leaves
Ron Kaufmann

Myoxanthus

Description: *Myoxanthus* orchids are distinguished by knobbed petals and bristly sheaths encasing the leaf stems. **Flower:** Small, fleshy flowers range from brown or maroon to green or cream, many with spots or stripes. Lateral sepals often are entirely fused, the dorsal sepal sometimes connected at the base. Narrower petals, frequently undulating, usually terminate in swollen, spoon-like tips. The small, faintly hairy lip, its edge often fringed or lobed, is hinged to the column base. Small wings or knobs decorate the short column. **Plant:** Tufted, creeping or climbing plants are composed of stems commonly longer than the narrow, leathery leaves. Stems are wrapped by brown, often papery, sheaths covered in bristles or stiff hairs. Clustered inflorescences emerge from the leaf base or lower stem, each bearing 1 flower. **Similar:** See *Dresslerella*. *Pleurothallis* lacks knobbed petals. *Restrepia* fused lateral sepals are shoehorn-shaped; all 3 sepals are knobbed.

Distribution and Diversity: Approximately 45 epiphytic, or uncommonly terrestrial, species flourish in a range of habitats and elevations. They can be found from Mexico to Bolivia and Brazil.

Ecology and History: The name *Myoxanthus* is taken from the Greek word for "muscle" (some say "doormouse"), possibly referring to the fleshy texture and purplish color. Most species are pollinated by small flies, often strongly attracted to such an appearance. Fly-pollinated flowers characteristically are brown or maroon, striped or spotted, and frequently present elongated, filament-like sepals or petals (see *Restrepia*). The knobbed tips of the sepals are odor-producing glands called *osmophores*: they emit musky or rancid odors some flies find irresistible, adding to the visual appeal provoked by their jittery motion in any breeze. The flowers of a few species (e.g., *M. cimex*) mimic insects and may deceive pollinators into attempting to copulate with the flower (see *Telipogon* and *Trichoceros*). The word "cimex" is Latin for bug, an example of a descriptive scientific name that highlights distinguishing features. Other examples include *M. punctatus* (spotted), *M. scandens* (climbing), *M. antennifer* (antenna-bearing), and *M. hirsuticaulis* (hairy stem). Some are more whimsical: *M. exasperatus*, for exasperated botanists stabbed by its spiky sheaths; and *M. uxorius* (meaning loving one's wife), named uxoriously by skilled taxonomist Carlyle Luer in honor of his wife, Jane.

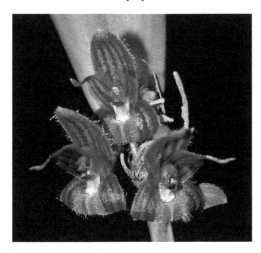

Myoxanthus ceratothallis
Andy Phillips

M. uxorius showing fused lateral sepals (below lip)
Andy Phillips

Myoxanthus cimex, the bug orchid, showing bristly sheaths
Andy Phillips

M. antennifer showing twin hair-like petal tips
Andy Phillips

M. exasperatus inflorescence
Andy Phillips

M. sarcodactylae
Andy Phillips

Myrmecophila

Description: Ant-inhabited pseudobulbs and showy flowers with broad lobes on the lip are unique to *Myrmecophila*. **Flower:** Large, eye-catching flowers vary in color, are long-lived, and often strongly fragrant. Sepals and petals are elongated, with undulating margins. Upward-curling side lobes of the broad lip are much larger than the middle lobe, which often is notched at the tip. The arching column is concave beneath, its sides narrowly winged. **Plant:** Swollen, cigar-shaped or conical pseudobulbs are hollow, with a small hole or slot in the base. Each bears several thick, fleshy, and short leaves. Exceptionally long inflorescences (more than 10' [3 m] in some species) arise from the pseudobulb apex, bearing flowers bunched at the top.

Distribution and Diversity: Approximately 10 rock-dwelling or epiphytic species favor deciduous forest, grasslands, and mangroves. They are found in the Caribbean and from Mexico to Venezuela.

Ecology and History: Large-bodied euglossine bees (*Eulaema*) fertilize flowers of *M. tibicinis*, picking up pollinia on their thorax when squeezing between the lip and column. Visits are concentrated in the early morning, when floral fragrance is strongest. Carpenter bees pollinate several other species. Bees that linger on the flower may be subject to attack by ants. *Myrmecophila* means "ant-loving," a reference to the hollow pseudobulbs' occupants. Colonies establish nests within the interior, where they lay eggs, rear offspring, and deliver food to larvae and adults. Ants enter the pseudobulb through a small opening, either provided by the plant or drilled by the ants themselves. This relationship is a classic mutualism: ants and orchid benefit from living together. The ants enjoy a safe nest location and devour nectar produced from bases of buds, flowers, and fruits. The orchid is defended by the aggressive ants against herbivores, such as flower-eating beetles, and even obtains nutrients from the colony's organic waste: food particles, dead kin, and other refuse discarded like compost inside the cavity. A variety of ants can participate in this mutualism, but only 1 species per plant. Smaller ants, however, provide less protection to their plants, which in turn suffer greater flower damage and reduced fruit set.

Myrmecophila tibicinis
Andy Phillips

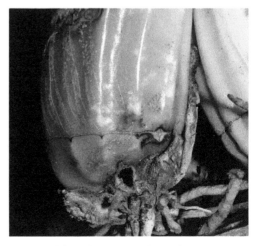

Myrmecophila hollow pseudobulb showing ant entrance
Ron Kaufmann

Myrmecophila albopurpurea showing upward curled
lip lobes
Andy Phillips

M. brysiana showing winged column
Andy Phillips

M. exaltata (orange color morph)
Andy Phillips

M. galeottiana
Andy Phillips

Notylia

Description: Gorgeous cascading showers of up to 150 small flowers characterize *Notylia*. **Flower:** Small (to 1/2" [1.3 cm]), delicate flowers are typically white and yellow, sometimes with darker spotting. Sepals are larger than petals, and lateral sepals often are fused, giving the flower a 4-part appearance. The distinctive lip is arrowhead-shaped and connected to the base of the column by a narrow strap. The column is unusual in that the anther cap covering the pollinia is located on the upper side of the column tip, giving a humped appearance; the genus name derives from the Greek word for "humpbacked." **Plant:** Small pseudobulbs, usually obscured in part by leafy sheaths, bear a single flat leaf (to 6" [15 cm] long). A densely flowered inflorescence dangles from the pseudobulb base.

Distribution and Diversity: More than 50 species of *Notylia* hang epiphytically in low-altitude wet forest. They can be enjoyed from Mexico to Brazil and Argentina.

Ecology and History: Like many orchids, *Notylia* flowers are pollinated by large euglossine bees (*Eulaema*) that visit to collect fragrances from the lip. These plants, however, display unique mechanisms to avoid self-fertilization and hybridization. The stigmatic area is located on the opposite side of the column from the hump enclosing the pollinia, reducing the risk of self-pollination. Furthermore, the flower's stigmatic area initially is too narrow to receive pollinia, but several hours after its own pollinia have been removed this receptive zone broadens and can be fertilized by a bee arriving from a different flower. Even if pollinia are not removed, the stigmatic zone still becomes receptive, but by then its own pollinia have become cemented in place, impossible to remove. This approach, in which the male parts of the flower are functional before the female parts, is known as *protandry*. Flower morphology also reduces the chance that pollen is transferred to a different species, a waste of reproductive material. Fine differences in column shape, lip orientation, and pollinia position ensure that a bee receives pollinia in a very particular location on his head, perfectly aligned with the stigmatic area of the same species but not with those of other, similar species.

Notylia trisepala showing arrowhead-shaped lip
Andy Phillips

N. lankesteri, after Charles Lankester (see *Chondrorhyncha*)
Franco Pupulin

N. pubescens inflorescence and leaves
Ron Kaufmann

Octomeria

Description: The combination of petite flowers emerging from a cluster of sheaths, atop reed-like stems wrapped in papery tubes, distinguishes *Octomeria*. **Flower:** Delicate, nearly translucent blossoms are pale in color, creamy white to yellow, occasionally with maroon striping. Rounded sepals and petals are alike, held open or in a vase shape; lateral sepals are fused in a few species. The lip usually is 3-lobed, commonly decorated with purple; side lobes often flank the diminutive column. Within the column are 8 pollinia, to which the genus owes its name. **Plant:** Lacking pseudobulbs, these small plants feature cane-like stems, each wrapped with a parchment-like covering and sporting a single leaf. Leaves are thick and fleshy, ranging from broad to aloe-like to cylindrical. The inflorescence arises from a compact cluster of woody or papery sheaths where the stem joins the leaf.

Distribution and Diversity: More than 150 species occupy a wide range of habitats as epiphytes and rock dwellers. The genus is found in the Caribbean and from Belize to Uruguay, with the greatest diversity in Brazil.

Ecology and History: *Octomeria* flowers attract pollinating flies with a combination of scent and nectar, in a system known as *myophily* (see *Bulbophyllum* and *Pleurothallis*). Flies generally are considered ineffective pollinators, because they are promiscuous, visiting many species and thus wasting pollen. Furthermore, flies seeking an egg-laying site often spend long periods exploring a single flower, a behavior that increases the risk of self-pollination. While many orchids rely on pre-pollination mechanisms to prevent selfing, such as attraction of specific insects by unique fragrance combinations, *Octomeria* relies on genetic incompatibility. Three species studied in Brazil (*O. crassifolia*, *O. grandiflora*, and *O. praestans*) showed virtually no fruit production when self-pollinated, but high rates of success (60–90%) when cross-pollinated. In selfed flowers, nearly half the pollen grains failed to develop fully, suggesting a biochemical incompatibility that blocks fertilization when pollination occurs. Another Brazilian species, *O. loddigesii* (synonymous with *O. graminifolia*), commemorates illustrious orchid grower George Loddiges (1786–1846), whose cavernous hothouses in England pioneered the use of misting systems to generate the consistent moisture needed to sustain his enormous displays of tropical plants.

Octomeria grandiflora
Ron Parsons

O. praestans showing
parchment-wrapped stems
Ron Kaufmann

O. juncifolia showing clustered
inflorescence sheaths
Andy Phillips

Oncidium

Description: The Dancing Lady flowers of *Oncidium* resemble a performer swirling her skirts, arms outstretched. **Flower:** Brilliant, usually yellow flowers are medium sized (1–2" [2.5–5 cm]) and commonly fragrant; a few are bright reddish purple. Similar sepals and petals are moderately narrow, often with brown bars or splotches. Perpendicular to the column, an elaborate lip combines a ruffled outer margin (the skirt) with outstretched lobes at the base (the arms). Between the lobes lies a warty hump, the source of the genus name: Greek for "wart-like." The short column has distinct lateral wings and a pad of tissue between the lip and stigmatic area. **Plant:** Vegetatively diverse, most species have shiny, flattened pseudobulbs yielding 1 to 2 leaves from the tip, and a pair of opposite, leaf-like sheaths at the base. Leaves are thin, often stiff, but highly variable in size. Inflorescences of 1 to 100 (or more) flowers arise from bulb sheaths and can be extremely long. **Similar:** See *Cyrtochilum* and *Gomesa*. *Trichocentrum* leaves are fleshier and much larger relative to its diminutive pseudobulbs.

Distribution and Diversity: This revised genus contains more than 500 species including many from previously well-known genera (e.g., *Odontoglossum*). Epiphytic and terrestrial, they grow in humid forests from low to high elevation in Florida and the Caribbean, and from Mexico to Bolivia, with 1 species in Brazil. Some authors place species with long, wiry rhizomes in *Otoglossum*.

Ecology and History: Some species are pollinated by diverse bees, particularly large species that collect oils from glands on the lip (see *Ornithocephalus*). Others (e.g., *O. hyphaematicum* and *O. planilabre*) take advantage of male *Centris* bees that patrol reproductive territories, savagely attacking any other male who enters. Their flowers, quivering in the breeze, mimic flying bees and are dive-bombed by the resident male. Striking with an accuracy of less than 1 mm, the deceived pugilist receives pollinia squarely on his forehead. The accurately named *O. abortivum*, and several other species, enhance their floral displays with aborted, sterile flowers. Inflorescences bear a few fertile flowers, alongside numerous partly developed blossoms lacking a functional column: these infertile flowers augment the mesmerizing visual presentation, and emit enticing scents. Such dazzling, confetti-like sprays of *Oncidium* were responsible in part for the origins of orchid mania in Europe (see *Cattleya*).

Oncidium henning-jensenii
Franco Pupulin

O. dichromaticum
Franco Pupulin

O. stenotis showing warty callus
Franco Pupulin

Oncidium hyphaematicum, column lobes and pad
Ron Kaufmann

O. hastilabium, a fragrant flower
Joe Meisel

O. lehmannii (formerly *Odontoglossum cristatellum*)
Ron Kaufmann

O. gayi (formerly *Odontoglossum helgae*)
Ron Kaufmann

Caucaea olivacea (formerly *Oncidium olivaceum*); white- and purple-flowered species from the high Andes are best treated as members of *Caucaea*
Ron Kaufmann

O. klotzschianum showing typical spray of flowers
Ron Kaufmann

Ornithocephalus

Description: Its name signifying "bird's head" in Greek, *Ornithocephalus* is distinguished by the column's elongated, beak-shaped tip, and leaves arranged in fans. **Flower:** Tiny flowers often are hairy, as is the inflorescence. Cupped or spreading sepals and petals are white to greenish yellow, sometimes with green stripes. The lip usually is 3-lobed and bears a thick, fleshy hump near the base. The green column is extended into a long, often undulated point, resembling an elephant's trunk or a stork's bill. **Plant:** Attractive fans of leathery, flattened leaves are joined by a seam to small sheaths at the base. Unbranched inflorescences arise from the sheaths. Roots mostly are hairy. **Similar:** See *Erycina* and *Macroclinium*.

Distribution and Diversity: Some 50 epiphytic species prefer humid habitats from low-elevation savanna to upper-elevation montane forest. Many species favor the dimly lit understory. The genus ranges from Mexico and Trinidad to Bolivia and Brazil.

Ecology and History: *Ornithocephalus* flowers attract bees with an open, oil-producing gland on the base of the lip. Small bees scramble up the lip to collect these fatty oils, shoving aside the column's long beak. The beak connects to a cap enclosing the pollinia, which are released when it is bent; a concave depression on the column's base receives the pollinia. Such oil-producing flowers are not uncommon in orchids (see *Gomesa* and *Rudolfiella*). They attract a very specialized group of bees, however, that constitute less than 2% of the world's bee species. Those pollinating *Ornithocephalus* collect the oil as a rich food source for their larval young: fats provide double the energy of carbohydrates. Oil rewards may have evolved in orchids as a more selective alternative to nectar. Myriad insects visit and rob nectar flowers of their reward, reducing the blossoms' attractiveness to true pollinators. In contrast, few insects can collect and utilize oils, and thus flower visitors are more likely to function as pollinators. Bees appear to have co-evolved with the orchids (and other oil-rewarding plants), developing a Swiss Army knife variety of brushes, blades, and other tools on their legs to gather and store the oil.

Ornithocephalus inflexus
Andy Phillips

O. cf. *gladiatus* showing beak-like column tip
Andy Phillips

Paphinia

Description: Showy flowers with an arrowhead-shaped lip decorated by frilly protuberances characterize *Paphinia*. **Flower:** Brownish-purple stripes or blotches overlie a cream or yellowish base. Sepals and petals are long and narrow, with pointed tips; petals are slightly smaller and darker. The 3-lobed lip, uppermost in these non-resupinate flowers, bears a tangled beard-like fringe on the outer region and fleshy fins on the middle. The column is long, arching, and winged near the tip. **Plant:** Egg-shaped pseudobulbs are smooth when young. Covered at the base by leafy bracts, each pseudobulb bears 2 to 3 leaves at the tip. Leaves are soft, long, and wide (to 10" [25 cm] by 2" [5 cm]), with clearly visible veins. Inflorescences hang from pseudobulb bases, displaying up to a dozen flowers. **Similar:** See *Houlletia. Sievekingia* has 1 leaf per pseudobulb, and leaves are thicker and waxier, not drooping.

Distribution and Diversity: Approximately 15 epiphytic species decorate wet low-altitude forest. They occur from Costa Rica to Peru and northern Brazil, and Trinidad.

Ecology and History: Like related genera, *Paphinia* is pollinated by euglossine bees. Male bees collect scent compounds used to attract females. Special foreleg brushes gather the fragrances, which are transferred to grooves on the hind legs. As the bee crawls out of the flower, his feet contact the sticky pollinia. The lip's beard likely forces the bee to walk on the column rather than flying away immediately. *Paphinia lindeniana* commemorates a celebrated explorer of Venezuelan orchids, Jean Linden (1817–1898). Only 19 years old when he left Belgium, Linden soon caught orchid fever, with an overwhelming interest in Andean species. He worked in inaccessible mountain locations and never revealed his exact collection sites, saying only that he brought orchids down "by roads which cannot be imagined by any who have not traversed them" (Reinikka 1995: 205). Linden made meticulous studies of the cool, breezy climate in these sites, recreating these conditions in his greenhouses. Before Linden, British gardeners grew all orchids, regardless of origin, in steamy coal-fired hothouses called *stoves*, a practice so misguided that England was labeled the "grave of tropical orchids" (Veitch 1889: 121). Linden changed all that, and cool weather orchids swiftly became popular.

Paphinia herrerae
Andy Phillips

P. rugosa showing fringed lip
Andy Phillips

P. neudeckeri showing lip fringe
and fins
Andy Phillips

Pelexia

Description: Terrestrial plants in *Pelexia* bear fuzzy inflorescences with distinctively hairy, tubular flowers. **Flower:** Dark green flowers are paler at the base and densely hairy. The dorsal sepal and petals converge in an elongated hood around the column. Narrow lateral sepals project forward and droop, or uncommonly, create a partial tube; they are fused at the base, forming a rounded bulge or pointed spur that is charged with nectar. A tiny, white lip bends downward sharply near the tip, its outer edge bristly, its base a flashy yellow. **Plant:** Dark green leaves with distinct stems are arranged in a rosette near the ground, often displaying white speckles or splotches of silver. A tall, upright inflorescence emerges from the middle of the leaves, with multiple flowers clustering near the top. **Similar:** See *Cyclopogon*, *Eltroplectris*, and *Gomphichis*. *Prescottia* flowers have a bonnet-shaped opening. *Sarcoglottis* flowers lack a visible spur.

Distribution and Diversity: Roughly 80 species grow terrestrially, occasionally epiphytically, in forest and damp grassland across a wide elevational range. They can be found from Florida, the Caribbean, and Mexico to Argentina and Uruguay.

Ecology and History: *Pelexia*, based on the Greek word for "helmet," sometimes is called the Helmet Orchid because of similarities between the flower and a medieval knight's steel headgear. *Pelexia oestrifera* and other species are pollinated by large bees (including *Bombus*), which sip nectar from the shallow spur. Flowers emit a musky odor that is detected by the keen olfaction of the bees. Because the compact flowers are small in comparison with these robust bees, the bees are obliged to grasp the sepals and lip while they shove their head into the tubular blossoms. The orientation of the column ensures that pollinia are attached to the bee's labrum, or upper lip. This is an advantageous placement, as the pollinia are not easily removed from the labrum and are protected during flight (see *Sarcoglottis*). These pollinators also forage on widespread, non-orchid flowers, suggesting that *Pelexia* nectar is sufficiently rich to divert the bees from more abundant food plants.

Pelexia congesta showing floral hood and nectar spur
Franco Pupulin

Pelexia inflorescence showing pale lip (upper left)
Franco Pupulin

Peristeria

Description: Round, bell-shaped flowers and large, smooth pseudobulbs topped by white, leafy sheaths are characteristic of *Peristeria*. **Flower:** Fleshy, medium-sized (to 2" [5 cm]) flowers are white or yellow with purplish speckling or dense blotches. Concave, waxy sepals are broad, curving forward with the smaller petals like a goblet. The hinged, 3-lobed lip curls down in front; large lateral lobes flare outward from the base. A beak-like projection points down from the tip of the short column. **Plant:** Densely clustered pseudobulbs are smooth and teardrop-shaped, often showing horizontal seams. Deeply pleated leaves arising from atop the pseudobulb can be very large (up to 4' [1.2 m] long). Short inflorescences (with one exception, *P. elata*), upright or hanging, emerge from the pseudobulb base and bear few to many flowers. **Similar:** See *Acineta* and *Lycomormium*.

Distribution and Diversity: Approximately a dozen epiphytic and terrestrial species prefer low-elevation sites on grassy edges and rock outcrops. The genus ranges from Costa Rica to Peru and Brazil.

Ecology and History: Large euglossine bees pollinate *Peristeria* flowers. Landing on the counterbalanced lip they crawl deeper into the cupped flower, at which point their weight tips the lip upward, throwing them against the underside of the column. Pollinia are attached to the thorax as the bee struggles to free itself. The erect lateral lobes help steer the bee as it enters. Many *Peristeria* species are fragrant. Where several occur together, they utilize distinct scents to attract only one type of bee, reducing the risk of hybridization. Where only 1 orchid species occurs, however, flowers emit a more general odor that appeals to a wide range of suitable bees. The national flower of Panama, *P. elata*, is known as the Dove Orchid for its brilliant white color and wing-like lip lobes. Unfortunately, plants grow in open, grassy areas where they are easily discovered and collected. Growth requirements are difficult to replicate and most wild plants perish in captivity. Rampant overcollection of this gorgeous orchid pushed the species to the brink of extinction. Fortunately, its declaration as the national flower has focused attention on its plight and prompted conservation efforts throughout Panama.

Peristeria elata, national flower of Panama
Ron Kaufmann

P. pendula
Andy Phillips

P. pendula showing pseudobulbs and seams (at top)
Ron Kaufmann

Pescatoria

Description: The large, clumping plants of *Pescatoria* are distinguished by a convex lip adorned with raised, rounded, parallel ridges. **Flower:** Fragrant flowers are creamy white, yellow, purple, or rich maroon. Thick sepals and petals are similar, often curling forward. The showy lip sometimes is covered with dense hairs, giving way to a series of keels radiating from the lip's base. A blunt column hovers above these ridges. **Plant:** Leaves arranged in loose fans are thin and veined (to 24" [61 cm] long), usually with an obvious joint near the bottom. Inflorescences from leaf bases support a single flower and 1 to 2 leafy bracts. **Similar:** See *Cochleanthes* and *Huntleya*.

Distribution and Diversity: Some 25 species of *Pescatoria* (aka *Pescatorea*), flourish as tree trunk epiphytes in low- to middle-altitude wet forests. They are found from Costa Rica to Peru and northern Brazil.

Ecology and History: *Pescatoria wallisii* and others are pollinated by euglossine bees (*Eulaema*). The raised keels guide the bee deep into the flower; backing out, it brushes the column and pollinia adhere to its back. The exact placement (on a dorsal plate called the *scutellum*) is typical for social bees that otherwise would rid themselves and their hive mates of pollinia (see *Xylobium*). *Pescatoria lehmannii* commemorates a dedicated explorer and assiduous collector of Ecuadorian and Colombian orchids, Friedrich Lehmann (1850–1903). He was appointed German Consul-General to Colombia, where he lived during the bloody, interminable War of a Thousand Days (1899–1902), wryly commenting, "Orchid collecting and travelling have been altogether hampered during the time of the civil war. Officially it has been pronounced ended over and over again, but while reading the announcement, if you were favorably situated, you would still hear the cracking of the rifles" (Cribb 2010: 28). He lived in Colombia for more than two decades, explored its orchid-rich forests widely, and confidently attributed his survival to two rules: "First, when in danger either from natives or, worse still, from lawless white men, I never produce a revolver or other weapon. Second, I never drink water without first boiling it" (ibid.). Sadly, water in its unboiled state was to be his ultimate demise, when he drowned in the Timbique River.

Pescatoria lehmannii showing hairy lip, and vegetation
Andy Phillips

P. klabochorum showing rounded lip keels
Andy Phillips

Phragmipedium

Description: *Phragmipedium* plants are known as Slipper Orchids for their flower's deeply pouched lip. **Flower:** Lateral sepals are fused and held beneath the lip; the dorsal sepal is broad and upright. Long, narrow petals droop down (to 36" [91 cm]), often twisting. The lip is clog- or slipper-shaped (but petal-like in *P. lindenii*), with the edges rolled inward. Flowers usually are green with a variety of brownish stripes and spots; a few species are brilliant red, pink, or purple. The short, bristly column is tipped with a triangular shield. **Plant:** Tall, arching leaves (to 36" [91 cm]) are slender, leathery, and grooved, growing in loose fans without appreciable stems. Tall inflorescences arise from leaf bases, bearing several flowers. **Similar:** *Selenipedium* has bamboo-like stems, pleated leaves.

Distribution and Diversity: Some 25 species dwell as terrestrials or rarely epiphytes, in low- to middle-elevation forests and grasslands. They occur from Mexico to Brazil and Bolivia.

Ecology and History: *Phragmipedium* plants have evolved archetypal trap flowers that capture bees and flies and provide them with only one exit: past the plant's sex organs. Pollinators slip on the in-rolled rim, plummeting into the pouch from which the waxy interior walls and cupped shape prevent escape. A stripe of bristles, however, provides a ladder up the back of the pouch. The lip's lateral lobes partly block the exit, forcing captives to clamber through a tiny opening beneath the column. There, the insect receives pollinia that it carries to another flower. Visual cues attract pollinators to the flowers: bee eyes are sensitive to shapes with a lot of edge, like the thin, twisting petals of *Phragmipedium*. Scent glands on the petal tips also provide olfactory enticement. Slipper orchids have uniquely rabid devotees, who traditionally have collected brightly colored Asian species. Their zeal propelled all slipper orchids onto the international endangered species list. Discovery of the vivid *P. besseae* and *P. kovachii* in tropical America unleashed a firestorm of overharvesting, virtually annihilating many wild populations. The latter orchid was smuggled illegally into the United States, earning Mr. Kovach and the Marie Selby Botanical Gardens (which named the species after him) criminal convictions and waves of negative publicity.

Phragmipedium besseae, in habitat (Ecuador)
Ron Kaufmann

P. kovachii
Valérie Léonard

Phragmipedium pearcei showing lip's deeply folded rear lobes
Andy Phillips

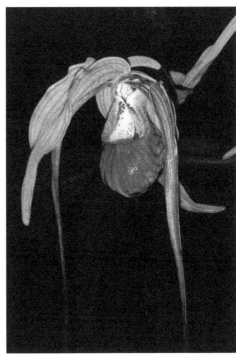

P. humboldtii
Andy Phillips

P. wallisii, with highly elongated petals
Andy Phillips

P. longifolium, in habitat
Ron Kaufmann

Platystele

Description: Although many genera of orchids can rightly be categorized as miniature, *Platystele* takes it down to a whole new level: they should be called microscopic orchids. **Flower:** Tiny, spreading sepals and petals are thin, translucent. Northern species tend to be orange; southern species are brownish red. The tongue-shaped lip is dark with a pale edge. Twin wings form a flat hood over the tip of the blunt column, the origin of the genus name (from the Greek for "broad column"). **Plant:** Diminutive plants lack pseudobulbs and form dense clumps or scrambling mats. Creeping stems are much shorter than leaves, which are leathery and rounded, with a central crease. The distinctive inflorescence of most species bears many flowers in 2 ranks, on opposite sides that open successively, leaving behind stubby, finger-like protuberances in a fishbone pattern. **Similar:** See *Brachionidium* and *Lepanthes. Pleurothallis* has stems longer than leaves and fused lateral sepals.

Distribution and Diversity: Nearly 100 species of epiphytic *Platystele* are sprinkled in cool, wet climates across a wide elevational range. They range from Trinidad and Mexico to Bolivia and Brazil.

Ecology and History: Some of the tiniest orchids in the world are in this genus, with evocative names like *P. microscopica, P. minimiflora,* and *P. microglossa.* Most flowers are less than a quarter of an inch across. Details of these petite but attractive blooms must be viewed with a magnifying glass or microscope to be appreciated properly. The smallest *Platystele* on Earth, recently discovered by accident in Ecuador, was found growing among the root mass of a larger plant: its flowers are only 2 mm across (<1/10th inch), and the gossamer petals are just 1 cell layer thick. The minute flowers of this genus rely on even more minuscule flies for their pollination. Nearly all species produce microscopic oil droplets on their flowers, attracting these pollinators. Small flies, such as fruit flies and fungus gnats, constitute a very diverse group of insects but are poor fliers, particularly in windy conditions. Their lack of mobility isolates *Platystele* populations, contributing to elevated rates of speciation.

Platystele showing hooded column
(center flowers)
Ron Kaufmann

P. umbellata
Andy Phillips

P. stenostachya showing fishbone
inflorescence
Andy Phillips

Plectrophora

Description: The combination of fleshy, blade-like leaves and trumpet-shaped flowers with a long nectar spur makes *Plectrophora* easy to recognize. **Flower:** Bright yellow to white flowers, often marked with spots and radiating stripes of burnt orange, can be quite large (to 2" [5 cm]) in some species. Sepals and petals, typically similar, are thrust forward or held open. The prominent lip forms a showy trumpet: long, tubular, with a curled rim. From the rear of the flower emerges an elongated, narrow, cylindrical spur; the genus name is taken from the Greek words for "spur bearing." **Plant:** Loosely clumped, small pseudobulbs, often obscured by leaf-like sheaths, each carry a single apical leaf. Fleshy, narrow leaves are flat or partially folded, lengthwise; they usually terminate in a dagger-like point (one species' name, *P. cultrifolia*, means "knife leaf"). Several inflorescences, each with 1 to few flowers, arise from the leaf bases. **Similar:** *Galeandra* pseudobulbs are tall, cylindrical.

Distribution and Diversity: Some 10 epiphytic species favor lower-altitude shady forest and scrub. They can be observed from Mexico to Bolivia and Brazil.

Ecology and History: The pale flower colors and long narrow nectar spur strongly suggest pollination by moths (see *Cryptocentrum*). Like many orchids, however, evidence from the field is lacking. It can truly be said that even 10,000 research projects on orchids would confirm only how much remains unknown. Sometimes entire species, not just their pollinators, evade discovery until someone looks in just the right place. *Plectrophora alata* was thought extinct in Mexico for more than 75 years; it had not been seen since being discovered in 1935, and its known range had been completely cleared for agriculture. Just 2 years after declaring the species extirpated, researchers surveying an area in Chiapas discovered several flourishing populations in rustic shade-coffee farms. The region had been severely deforested and only meager habitat patches remained. But coffee grown the traditional way, under a canopy of shade trees (*Inga*), provides surprisingly good habitat (see *Scaphyglottis*): *P. alata* thrives in the thick moss that cloaks the *Inga* trunks and coffee bushes.

Plectrophora triquetra showing entry to nectar tube
Andy Phillips

P. alata
Andy Phillips

P. schmidtii showing folded, pointed leaf
Andy Phillips

Pleurothallis

Description: This stupendously diverse genus, among the most speciose in tropical America, nearly defies description; however, *Pleurothallis* can be summarized coarsely as having small, 4-part flowers and leaf stems wrapped by ribbed sheaths. **Flower:** Lateral sepals are fused, totally or partially, and resemble the dorsal sepal. Smaller, usually narrower petals stretch out perpendicularly; the resulting flower appears 4-parted. The tiny lip is variably shaped, commonly tongue-like, and is hinged to the short column. Colors span the spectrum, but purples and browns are most common, as is the presence of spots and stripes. **Plant:** Densely clustered stems bear 1 leaf each, the stems partly encased in dry, corrugated sheaths: *Pleurothallis* literally means "ribbed stem." Leaves are variable, usually leathery and broad or notched at the base. Inflorescences arise from the leaf base, bearing 1 to many flowers, often lying atop the leaf blade. **Similar:** See *Brachionidium, Dryadella, Lepanthes,* and *Platystele. Stelis* flowers have tiny petals and a strongly triangular shape in most species.

Distribution and Diversity: At present, some 600 species reside in *Pleurothallis*; genetic analyses continue to redefine the genus, moving some species into new genera (e.g., *Acianthera, Specklinia*). Plants grow epiphytically or terrestrially, in low- to high-elevation forests in the Caribbean and from Mexico to Argentina.

Ecology and History: Quintessential fly-pollinated flowers dominate *Pleurothallis*. Various permutations of purplish flowers, speckles and stripes, dangling hairs wriggling in the breeze, and a faint rotting odor all catch the attention of flies. They arrive searching for nectar (a pollination syndrome known as *myophily*) or egg-laying sites (*sapromyophily*; see *Bulbophyllum* for both). Flies crawl up the hinged lip, their weight causing it to tip upward and press the unsuspecting insect against the column to receive, or deposit, pollinia. Fly-pollination syndromes have proliferated in a number of related genera, including *Dracula, Lepanthes, Masdevallia, Restrepia,* and *Stelis*. By and large these are orchids of higher elevations and cool climates, where the bees that typically pollinate lowland orchids decline in abundance. Bees are chiefly replaced by flies, a transition that has induced cool-growing orchids to evolve novel enticements, as flies respond to cues wholly different from those that stimulate bees.

Pleurothallis cordata
Ron Kaufmann

P. cf. *ortegae*
Andy Phillips

P. ascera showing leaf-top flower
placement
Andy Phillips

Pleurothallis with markings and bristles to attract flies
Joe Meisel

P. lynniana displaying stripes to entice fly pollinators
Ron Kaufmann

P. medinae
Andy Phillips

P. (Acianthera) ramosa showing thick, fleshy leaves
Ron Kaufmann

P. cf. *crocodiliceps*, example of white cross flower
Ron Kaufmann

Huge *P. gargantua* flowers reach 3" (7.5 cm) in height
Ron Kaufmann

Clustered *P. teaguei* flowers arising from atop a leaf
Ron Kaufmann

Distinctive bell-shaped flowers of *P. truncata*
Ron Kaufmann

Pleurothallis rowleei, example of hanging inflorescence
Andy Phillips

P. cf. *loranthophylla* hanging flowers
Ron Kaufmann

Epiphytic *Pleurothallis* growth showing stem sheaths
Ron Kaufmann

P. dunstervillei
Andy Phillips

Polycycnis

Description: *Polycycnis* derives from Greek words for "many swans," in reference to the numerous flowers whose lithe, arching columns resemble the necks of these graceful birds. **Flower:** Slender, pointed petals and sepals are yellow to tan with brownish-red blotches, spreading or more commonly drooping. The lip's outer section is densely hairy and narrowly joined to an inner section with flaring side lobes. The column is exceptionally long and curving, with a swollen tip (the swan's head). Most species are fragrant. **Plant:** Small, clustered pseudobulbs bear a single leaf from the apex and a hanging inflorescence from the base. Large leaves (to 16" [41 cm] long) are soft and heavily veined, often distinctly. **Similar:** See *Cycnoches*.

Distribution and Diversity: Approximately 20 epiphytic, rarely terrestrial, species embellish dense forest or sunlit hillsides at low altitudes. They can be enjoyed from Costa Rica to Peru and Brazil.

Ecology and History: Along with close relatives (*Coryanthes, Houlletia, Stanhopea,* and others), *Polycycnis* flowers attract male euglossine bees that gather scent compounds from the lip. Landing on the outer lip, his weight causes the entire flower to sag. The bee then climbs up to the lip's base where his forelegs scratch the surface for odors. He must take flight to transfer these odors to his hind legs, and in so doing he bumps the arching column now positioned directly behind him. Sliding past the column's tip triggers the release of pollinia, which are then glued to the last segment of the bee's back. After departing, euglossine bees utilize the odors in mating displays, to attract females and to battle rival males. Scientists employ these odors, too, as bait for bee traps. Many of the associations between bees and orchids cited in this book rely not on observations of a bee visiting a flower, but rather on the discovery of identifiable pollinia on the back of a captured bee. Any one species of orchid typically will mix several scent compounds to create its own unique fragrance (see *Gongora*). Flowers of *Polycycnis* exude at least four major scent classes: eugenol (odor of clove), cineole (eucalyptus), limonene (citrus), and sabinene (the spiciness of black pepper).

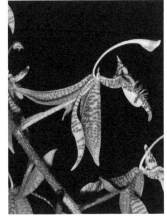

Bolivian *Polycycnis* cf. *silvana* showing hooded column
Andy Phillips

Polycycnis flower showing flared lip lobes
Andy Phillips

P. muscifera
Andy Phillips

Polystachya

Description: The small flowers of *Polystachya*, their lips characteristically bumpy or hairy, are packed tightly onto a spike like a stalk of wheat. **Flower:** The non-resupinate, nodding flowers can resemble a person wearing a Dutch hat, the lateral sepals drooping to form the hat flaps. Petals are smaller but similarly colored: yellowish green, pink, or white. The short column is held beneath a fleshy, undulating, tongue-like lip covered with mealy filaments or scales. **Plant:** Small pseudobulbs shaped like garlic bulbs, or occasionally like cigars, are usually sheathed by leaf bases. Two to 4 short leaves (to 8" [20 cm]), leathery and often notched at their tips, arise from the pseudobulb. The arched, sometimes branching, inflorescence emerges from the pseudobulb apex, bearing diminutive flowers and buds arranged closely together, like small ears of corn (*Polystachya* stems from the Greek for "many grains"). **Similar:** See *Bulbophyllum*. *Prosthechea* flowers are larger, less densely packed.

Distribution and Diversity: More than 200 mostly epiphytic species inhabit upper wet forest and grasslands. Their worldwide distribution is centered in Africa, with only a handful of species in the western hemisphere, where they range from Florida, the Caribbean, and Mexico to Brazil.

Ecology and History: *Polystachya* flowers use a rare strategy to attract small bees: they offer false pollen as a reward. Such pseudopollen has been found in few other New World genera (see *Maxillaria* and *Scuticaria*). Nearly all orchids have inedible pollinia rather than the nutrient-rich powdery pollen common in other flowering plants. Some orchids offer nectar as a reward, although a third of all species offer no reward at all. In *Polystachya*, which provide no nectar, the mealy hairs of the lip are actually tiny columns of epidermal cells, only a few cells tall. These special, glandular cells are rich in protein and starch. Female halictid bees visit species such as *P. concreta*, drawn by scents or visual cues. They collect the tips of the hairs, which break off into a dusty, pollen-like substance. The bees store the pseudopollen in brushy areas on their hind legs for later consumption. As the bee squirms out of the flower its thorax drags over the column and pollinia are attached or deposited.

Polystachya masayensis showing mealy scales on lips
Franco Pupulin

P. lineata showing drooping sepals
Franco Pupulin

Ponthieva

Description: *Ponthieva* ("pon-THEE-va") orchids favor damp sites and are distinguished by a non-resupinate (upside-down) flower orientation and partly joined petals forming a lowermost pseudo-lip. **Flower:** Pale white or greenish flowers, often marked with reddish or brown spotting or veins, are borne on an upright, unbranched spike. Sepals typically are broad and spreading, occasionally hairy on the edges. The 2 petals are connected at least at the base and fused to the column; these petals lie atop the central sepal, providing a landing platform for insects. The small lip is uppermost, attached to the base of a sturdy column, and is slightly to deeply concave. **Plant:** A whorl of soft leaves, occasionally hairy or weakly pleated, lie nearly flat on the ground. Pseudobulbs are absent.

Distribution and Diversity: Nearly 60 terrestrial species lurk in wet sites from low to high elevations; a few are epiphytic. They can be found from the southeastern United States through the Caribbean to Chile and Argentina.

Ecology and History: Several species (e.g., *P. racemosa, P. ventricosa*) are called the Shadow Witch: similar monikers have been given to terrestrial orchids that appear as if by magic to flower in the early spring. Alternatively, the name may reflect the former use of the flower in witches' love potions; vials of Shadow Witch perfume still can be obtained from obscure fragrance dealers. The plants appear to be pollinated by flies and bees and have evolved curious strategies to attract these insects. Oil glands in the lip produce a reward for visitors, which land on the pseudo-lip and walk over the top of the column to reach the erect lip: along the way they pick up pollinia on their mouthparts and deposit them on the next flower. Like many fly-pollinated orchids, the combination of dark spotting and fringing hairs on the flowers renders them visually appealing; several species even resemble flying insects. As a final trick, *P. maculata* sepals are decorated with glittering dots resembling droplets of honey. The dots, in reality hard structures that brilliantly reflect light, further pique the interest of passing bugs.

Ponthieva tunguraguae showing "pseudo-lip" of fused petals
Andy Phillips

P. tuerckheimii with tiny yellow pollinarium
Andy Phillips

P. diptera inflorescence
Ron Kaufmann

Porroglossum

Description: The trap flowers of *Porroglossum* are surprisingly fast-acting, their sensitive lip closing in seconds when touched. **Flower:** Sepals, partly fused into a cup, bear tips variously elongated into blunt lobes or trailing, thread-like tails. The dorsal sepal is concave. Petals are much smaller, thus flowers appear three 3-parted. Blossoms may be oriented with the tongue or spoon-shaped lip above, or below, the column. Its base is narrow and wraps around the foot of the column like a hinge. **Plant:** Clustered or creeping small plants (3–4" [7.5–10 cm] tall) lack pseudobulbs. Leaves are distinctly stalked, their surface often covered in warty bumps. The inflorescence, sometimes bristly or warty itself, is much longer (to 6" [15 cm]) than the leaves and yields small flowers (to 1" [2.5]) that open one at a time.

Distribution and Diversity: Nearly 40 species thrive in cool, mountainous forest as epiphytes and terrestrials. The genus ranges from Venezuela to Bolivia, its center of diversity in Ecuador.

Ecology and History: Like many close relatives, *Porroglossum* flowers are pollinated by small flies. The behavior of flies on orchid flowers is erratic, and they are more unreliable as pollinators than bees (see *Octomeria*). Fly-pollinated orchids utilize a variety of techniques to ensure these fitful visitors successfully contact the column and fertilize the flower, but none so elaborate as in *Porroglossum*. Alighting on the lip, a fly's touch, no matter how faint, triggers a mousetrap-like mechanism that snaps the lip against the column. The unique connection between the base of the lip and the column is the key. The lip's strap-like hinge is bent elastically over the column foot, awaiting release. Minuscule changes in the water pressure of cells in this hinge are all that is needed to unleash the spring-loaded lip, which clamps against the column in less than 2 seconds. The hapless fly is trapped and can escape only by wriggling between the lip and petals: in the process, pollinia are glued to its back. Once triggered, the lip automatically resets in about 30 minutes: the water pressure changes are reversed, and the mousetrap is set for the next victim. Many species close naturally at night, denying access to non-pollinating nocturnal insects.

 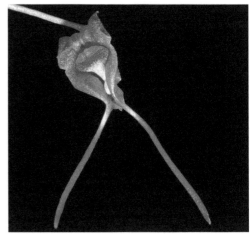

Porroglossum amethystinum, lip open
Andy Phillips

P. amethystinum, lip closed
Andy Phillips

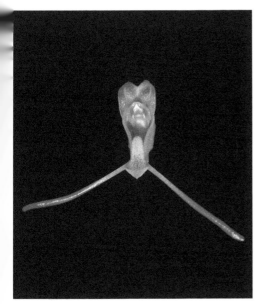

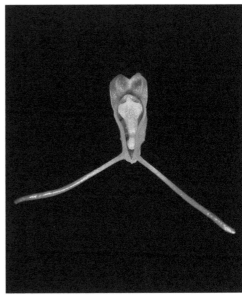

Porroglossum aureum
Andy Phillips

P. aureum, lip closed
Andy Phillips

P. josei, showing tiny, lobe-shaped petals
Andy Phillips

P. josei, lip closed
Andy Phillips

Prescottia

Description: *Prescottia* flowers ought to be called Bonnet Orchids, for the shape of the narrow entrance to their deeply pouched lip. **Flower:** Strongly fragrant, the tiny flowers are packed onto a vertical spike, each oriented lip-uppermost (non-resupinate). Small sepals and petals are white or pink and often curl backward so strongly as to be rolled up. Sepals are joined at the base, forming a sort of crown atop which the lip reposes. The green, fleshy lip is profoundly concave, almost ladle-shaped, with a bonnet-like opening. **Plant:** Plump roots give rise to a whorl of variably shaped leaves, either stalked or narrowed at the base, that are shed annually. **Similar:** See *Gomphichis* and *Pelexia*.

Distribution and Diversity: Some 25 terrestrial species favor cool, middle- to high-altitude woodlands and wet grasslands. The genus ranges from Florida, the Caribbean, and Mexico to Paraguay and Uruguay.

Ecology and History: Bees and moths largely pollinate *Prescottia* flowers, alerted by their potent fragrance to the presence of nectar secreted by glands on the lip. To reach the flower's interior, the insects mash their heads into the bonnet's narrow opening, and in so doing nudge the column and pick up pollinia on their face or mouthparts. *Prescottia stachyodes*, which releases a spicy-sweet odor after dark, is visited by small pyralid moths. This flower demonstrates protandry: development of male reproductive organs before female organs. For a flower's first 3 days the column bends toward the lip wall, prohibiting access to the stigmatic region; later the column straightens to point at the lip opening, allowing the stigma to receive pollinia. Because flowers on the spiked inflorescence develop from the bottom toward the top, the lowermost flowers tend to be older and therefore in the female phase, while the uppermost, younger flowers are in the male phase. This is common in many terrestrial orchids and is considered an adaptation to bumblebee pollination. Bumblebees habitually ascend, rather than descend, flowering spikes. As long as male phase flowers are at the top, pollinia are not passed to lower female flowers, and the bee is more likely to fly away and cross-pollinate another flower.

Prescottia cf. *villenarum* showing rolled petals and sepals
Andy Phillips

P. aff. *stachyodes* showing bonnet-like lip opening
Andy Phillips

Prescottia terrestrial vegetation, in habitat
Ron Kaufmann

Promenaea

Description: *Promenaea* are characterized by showy yellow or cream flowers, often patterned with purple, and by soft leaves with visible veins. **Flower:** Yellow, pale green, or cream colors dominate medium-sized flowers (1–2" [2.5–5 cm]); most species also bear burgundy speckling or tiger striping. Waxy sepals are spreading or slightly cupped and larger than petals. The forward-jutting lip is overtly 3-lobed, with erect wings flanking the column, between which stretches an undulating, fleshy ridge. Many species are strongly fragrant. **Plant:** Clustered pseudobulbs of these diminutive plants are short (1" [2.5 cm]) and somewhat 4-sided. Several soft leaves with evident veins emerge from atop the pseudobulb. An arching or pendant, short inflorescence emerges from the base, bearing 1 to 2 flowers. **Similar:** *Zygopetalum* flowers have a ribbed, yoke-like collar across the lip's base.

Distribution and Diversity: Nearly 20 species of *Promenaea* gild lower- to middle-elevations as epiphytes or in rocky crevices. They occur only in the moist forests of southern and central Brazil.

Ecology and History: *Promenaea* flowers are pollinated by medium-sized, male euglossine bees attracted to the floral perfume. The side lobes of the lip guide the bee, like twin velvet ropes, beneath the column to assure pollination. Unfertilized flowers remain open for many weeks, prolonging the plant's attractiveness. As a general rule, tropical plants hold their flowers open for short periods only, because pollinators typically are abundant. Doing otherwise would be a waste of precious energy. Orchids that rely on highly specialized pollinators, however, are the exception to this rule. They must invest energy to maintain an attention-grabbing display, whether visual or olfactory, until the right bee can discover and pollinate a flower. In *Promenaea* an added twist occurs: the bright flower colors fade swiftly after pollination and are replaced with green. This "re-greening" involves the revival of chloroplasts that become photosynthetically active again. The adaptation may have arisen to capture more sunlight and pay back the energy invested in floral longevity. The more immediate benefit lies in improved pollination efficiency, as the visual signal directs bees away from fertilized flowers toward receptive blooms.

Promenaea guttata showing lip wings and ridge between
Andy Phillips

P. xanthina
Andy Phillips

Prosthechea

Description: Elongated pseudobulbs and fleshy flowers with partial fusion of the lip and column distinguish *Prosthechea*. **Flower:** Typically plastic or waxy in texture, the large (to 5" [12.7 cm] across) flowers are yellow or greenish yellow, but reddish orange in a few species. Sepals and petals are similar, and most species have flowers oriented lip-uppermost (non-resupinate), the lip hooding the column and fused to it for half the lip's length. The tip of the column bears small ears, and a knob-like bump from which the genus takes its name, Greek for "appendage" (and the root of "prosthesis"). Many species have pleasantly fragrant flowers. **Plant:** Mostly clustered pseudobulbs range in shape from eggs to slender eggplants. Strap-like leaves, leathery but thin, emerge from each pseudobulb, separated by a thin seam. A distinctive leafy bract clasps the base of the inflorescence. **Similar:** See *Polystachya*. *Encyclia* flowers are oriented lip-lowermost, the lip joined only to the column base, and pseudobulbs are dissimilar.

Distribution and Diversity: More than 100 species of *Prosthechea* reside as epiphytes and rock dwellers in seasonal forest and scrub. They occur across a wide altitudinal range, from southern Texas and the Caribbean to Venezuela, Bolivia, and Brazil.

Ecology and History: The partial fusion of the lip and column forces the pollinators, medium-sized wasps, to squeeze beneath the column to reach the scented base of the lip. In doing so they fertilize the flower. One species, *P. vitellina*, is pollinated by hummingbirds that respond to its bright orange color (wasps cannot see orange or red): the bill is forced beneath the column, and pollinia are attached to the bird's forehead or bill. *Prosthechea cochleata* is the national flower of Belize, where its dark purple lip markings earned it the local name Black Orchid; however, no truly black orchids exist in nature. Several species in this genus have proven medically useful. In Mexico, *P. michuacana* and *P. varicosa* are called *camote de agua*, or water bulb: chewing the pseudobulbs was a common traveler's trick to slake their thirst. Extracts from the former species recently were proven to promote wound healing and reduce inflammation, a practical example of the value of biodiversity.

Prosthechea crassilabia
Ron Kaufmann

P. vitellina, pollinated by hummingbirds
Andy Phillips

Prosthechea cochleata, national flower of Belize
Andy Phillips

P. fragrans showing column knob and ears
Andy Phillips

P. cf. *fragrans* showing pseudobulbs and inflorescence
sheaths
Ron Kaufmann

Prosthechea inflorescence and leaf
Ron Kaufmann

Pseudolaelia

Description: *Pseudolaelia* plants, restricted to eastern Brazil, have *Laelia*-like flowers, but rambling growth forms and loosely spaced pseudobulbs. **Flower:** Moderately narrow sepals and petals are pink to lilac, rarely yellow or white. The flared front lobe of the lip is ruffled, matching the petals in color; a wide, contrasting streak marks the center. Slender side lobes are held outstretched like arms from the lip's base. The column tip expands into broad, paired wings. **Plant:** Spindle-shaped pseudobulbs, with multiple seams, are sparsely distributed along a scrambling rhizome. Several narrow leaves sprout from each pseudobulb, while dry bracts surround the base. Leaves are leathery, or thin and visibly veined; they may be shed during dry seasons. A long, slender inflorescence emerges from the leaf bases, bearing few to many flowers. **Similar:** See *Laelia*. *Rhyncholaelia* is restricted to Central America.

Distribution and Diversity: At least 18 species are found only in dry, eastern Brazil. They occur in coastal habitats and on granite outcroppings, as rock dwellers or epiphytes.

Ecology and History: *Pseudolaelia* flowers exhibit "invisible" stripes advertising a pathway to nectar: the stripes can be seen only in the ultraviolet spectrum, visible to bees but not humans. Such nectar guides are common in bee-pollinated flowers. Studies of *P. corcovadensis* revealed they are deceit mimics, proffering striped flowers but no nectar. If bumblebees (*Bombus*) are not successfully lured, the flower can self-pollinate. *Pseudolaelia* epiphytes grow almost exclusively on strange, gnarled *Vellozia* shrubs. One species is even named *P. vellozicola*. The orchid's scrambling growth allows it to climb the twisting *Vellozia* branches readily. The shrubs and *Pseudolaelia* species are among the scant hardy plants that survive atop steep-sided granite mountains known as *inselbergs*. These exposed rock summits offer no shelter from sun, wind, and desiccation. *Vellozia* plants form compact bunches or mats to better shade and protect scarce, humid soil. Like *Pseudolaelia*, its roots are sheathed with velamen (see *Huntleya*) that quickly soaks up meager moisture. *Vellozia* is a true drought tolerator, among only 300 plants worldwide that can suspend photosynthesis and enter dormancy to withstand a near total loss of moisture; at the first rainfall, however, it swiftly revives.

Pseudolaelia calimaniorum
Ron Kaufmann

P. corcovadensis showing column
wings
Ron Kaufmann

P. vellozicola pseudobulbs and
leaves on *Vellozia* shrub
Ron Kaufmann

Psychopsis

Description: Commonly called Butterfly Orchids, the arresting flowers of *Psychopsis* are readily recognized by the wing-like, ruffled lateral sepals and the antennae-shaped petals and dorsal sepal. **Flower:** Marked with tiger striping of reddish brown on yellow, the stunning flowers are large (to 6" [15 cm] across) and appealing. Upper floral elements are long and ribbon-like, while the lateral sepals are broader, curved, and evince wavy edges. The flaring lip, also deeply ruffled, is yellow at the center with irregular brownish markings around the rim. Small, rounded side lobes of the lip stand adjacent to the short column, between them a fleshy hump. **Plant:** Clumped pseudobulbs are flattened laterally, each bearing a single leaf, and wrapped by opposing sheaths. Leathery leaves usually are mottled or spotted. Flowers are borne on long (to 18" [46 cm] or more), thickly jointed inflorescences, with just a single blossom open at a time. **Similar:** *Oncidium* petals and dorsal sepal are not thread-like.

Distribution and Diversity: Four or 5 species epiphytically adorn low-elevation, wet forest. The genus occurs in Trinidad, and from Costa Rica to Peru and Brazil.

Ecology and History: In ancient Greece, Psyche, from which the name *Psychopsis* is taken, meant "butterfly"; only later did it come to refer to the mind or soul. Certainly these lovely flowers resemble brilliantly colored butterflies, especially when presented singly on a long, dangling stem, fluttering in the breeze with their ersatz antennae dancing lightly. The similarity is so close that many have suggested the flowers may attract inexperienced male butterflies that attempt either to mate with the flower (pseudocopulation) or attack it (pseudoantagonism). Orchid expert Calaway Dodson (see *Stenia*) reported seeing zebra butterflies (*Heliconius*) attack *Psychopsis* flowers in the 1950s; however, despite their broad range and singular attractiveness no research has definitively confirmed the pollinator's identity. The enchanting flowers of *P. papilio* were reputed to have caught the interest of the Duke of Devonshire (see *Galeandra*), providing the initial spark for his lifelong dedication to orchids. Sadly, the allure of *Psychopsis* flowers has led to virtual extinction through overcollection for species in Costa Rica (*P. krameriana*) and Venezuela (*P. papilio*).

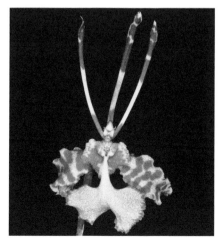

Psychopsis papilio
Ron Parsons

P. krameriana
Ron Parsons

Restrepia

Description: Attractive *Restrepia* flowers are easily identified by their distinct shape, like a colorful shoehorn with 3 antennae. **Flower:** The petals and dorsal sepal are the antennae, each membranous at the base but largely threadlike, ending in a knob. The shoehorn is not the lip, but rather the oblong, fused lateral sepals, occasionally notched at the tip. The lip itself is inconspicuous, but decorated with 2 hairlike projections. Flowers are reddish, green, or yellow to orange, often with red or purple striping or spotting, especially on the shoehorn. Most species have peculiar or rank-smelling fragrances. **Plant**: Pseudobulbs are absent. Plants form clumps of short stems covered loosely by papery sheaths, with fleshy, teardrop-shaped leaves; the genus is readily diagnosed by these features alone. Inflorescences often arise from the backs of leaves, producing 1 to few flowers. **Similar:** See *Myoxanthus*.

Distribution and Diversity: More than 50 species are all quite distinct from other genera, but many species display considerable variability and no two populations are alike. Largely epiphytic, and more common on tree branches near water, plants favor middle- to high-altitude moist forests, rarely lowlands. They range from Central America to Bolivia, with peak diversity in the Andes.

Ecology and History: *Restrepia* flowers are fly pollinated and show many of the unusual adaptations that attract flies. The thin, antennae-like dorsal sepal and petals wave slightly in the breeze, a motion known to beckon insects. More important, the clubbed tips of these structures contain small glands, or osmophores, that release foul-smelling fragrances designed to lure flies. The osmophores first excrete energy-rich, fatty oils, which in turn power synthesis of complex and far-carrying perfumes. This two-stage process may begin even while the flower is still in bud. A similar pathway is found in mint plants (*Mentha*), although their sweet perfume appeals to bees and hummingbirds, and in insectivorous plants. *Restrepia* flowers offer no nectar reward, and do not devour their visitors; their fragrances are highly volatile, however, wafting rapidly and over great distances to attract pollinators. Flies following the enticing scent discover only a putrid-smelling flower, not a food source, at the end of their search.

Restrepia sanguinea
Andy Phillips

R. antennifera
Andy Phillips

R. guttulata
Andy Phillips

R. brachypus, leaves and
pale stem sheaths
Ron Kaufmann

Rhyncholaelia

Description: The 2 species of *Rhyncholaelia* have white, *Cattleya*-like flowers, and a peculiar beak on the seed capsule's tip (*rhyncho* is the Greek word for "beak"). **Flower:** Pale green sepals and petals are relatively narrow, while the white or green lip is broad. The base of the lip wraps the column and is colored with emerald or rose. *Rhyncholaelia digbyana* displays a flamboyantly frilly lip margin, whereas *R. glauca* has a smooth lip. **Plant:** Narrow pseudobulbs are elongated and slightly flattened, often obscured by white sheaths. Each bears a single, grayish-green leaf with a waxy surface and faintly pointed tip. Inflorescences emerge from the pseudobulb apex, producing 1 flower each. **Similar:** See *Brassavola* from which these species were moved; also *Cattleya*, *Guarianthe*, and *Pseudolaelia*.

Distribution and Diversity: The 2 known species are epiphytes in seasonally dry forests at low to middle elevation. They can be seen from Mexico to Nicaragua.

Ecology and History: White flowers with long nectaries and strong fragrance are excellent examples of the moth pollination syndrome (see *Brassavola*). These flowers harbor a tubular nectary, hidden within the stem (pedicel), which contains sweet liquid for the moths. The pale blooms and powerful perfume jump out like a neon sign to a night-flying moth. In Mexico, populations of *R. glauca* flower synchronously, and all plants hold their blossoms open for nearly 3 weeks. When pollinated, however, an individual flower closes almost immediately, encouraging subsequent moths to visit other, unfertilized flowers. In Honduras, the entrancing *R. digbyana* was selected as the national flower. It has been widely popular there, not only as an ornamental plant but also for medicinal purposes: the sap reputedly can be used to stanch hemorrhages. Conservation and education measures have been put in place to ensure that the Honduran national flower is protected for future generations. Both species have been used extensively in breeding commercial orchid hybrids. In particular, *R. digbyana* has imparted its frilly lip and sweet, citrus-like fragrance to many varieties of showy, large-flowered *Cattleya* hybrids. Production of a single flower is genetically dominant, however, and most *Rhyncholaelia* hybrids bear few flowers per inflorescence.

Rhyncholaelia digbyana, national flower of Honduras
Andy Phillips

R. glauca showing lip wrapping column
Andy Phillips

Rodriguezia

Description: Pale flowers with short, downward-curving nectar spurs are characteristic of *Rodriguezia*. **Flower:** Petals and dorsal sepal sweep forward, while lateral sepals connect in a boat-like shape below. The lip is long, shallowly notched, and bears a short nectar-producing horn at its base. Flowers are white or pink, with yellow and red markings on the lip. The column is thin and decorated with twin arm-like projections. **Plant:** Small, egg-shaped pseudobulbs, their bases clasped by leafy sheaths, are connected by a short, or distinctly elongated, horizontal stem. One or 2 leaves (to 9" [23 cm] long) arise from each pseudobulb, as do unusually long, twining roots. Leaves are leathery, typically folded near the base. A short inflorescence emerges from the pseudobulb sheaths, often bearing flowers on only one side. **Similar:** See *Comparettia* and *Gomesa*.

Distribution and Diversity: Nearly 50 species of *Rodriguezia* grow as twig epiphytes in humid forests from low to middle elevation. They range from Mexico to Peru and Brazil.

Ecology and History: Most species are pollinated by euglossine bees, although several are visited by hummingbirds. In *R. lanceolata*, the column arms guide the hummingbird's beak into the flower, like a sword into a sheath, ensuring that pollinia are affixed to its base. *Rodriguezia* flowers have pale or gray pollinia, quite different from the standard bright yellow, reducing the likelihood that the hummingbird will notice and remove them (see *Elleanthus*). *Rodriguezia* plants typically grow on small twigs and show many adaptations to this tenuous lifestyle. Twining roots and long stems hold plants in place, and rapid development allows them to flower before twigs can snap. They utilize an alternative form of photosynthesis, known as CAM (see *Macroclinium*), which reduces water loss fatal to small plants growing far above a reliable supply of moisture (see *Ornithocephalus*). These and other adaptations have permitted orchids to colonize ubiquitous but challenging twig sites, opening the door to an explosion of new species. DNA evidence now suggests that several genera of twig epiphytes have contributed more than any other group to the enormous worldwide diversity of orchids. In the modern era, more than 70% of all orchid species, including a plethora of twig specialists, are epiphytic.

Rodriguezia batemanii
Andy Phillips

R. venusta; part of nectar spur (upper right flower)
Andy Phillips

R. lanceolata in habitat, showing twining roots and foliage
Ron Kaufmann

Rudolfiella

Description: *Rudolfiella* can be recognized by the combination of pleated leaves and vivid flowers featuring a lip with broad lateral lobes and a slender base. **Flower:** Showy flowers are bright yellow with rich brown speckles and blotches. The dorsal sepal stands upright, while lateral sepals are joined at the base to each other and the column, forming a fleshy chin. Shorter petals typically project forward. The broad lip spreads a pair of large, wing-like side lobes, then narrows drastically to a stalked attachment. Between the lobes lies a fleshy hump (callus), often covered in short hairs, directly beneath the curved column. **Plant:** Spherical or egg-shaped pseudobulbs often are compressed, each producing 1 wide, deeply pleated leaf with prominent veins. Long, arching inflorescences bear numerous flowers. **Similar:** See *Bifrenaria*.

Distribution and Diversity: Just 6 epiphytic species of *Rudolfiella* brighten low altitude, warm and humid forests. Their center lies in Brazil, but they range from Panama to Bolivia.

Ecology and History: Bees are the presumed pollinators of this small but colorful genus. In one species, *R. picta*, specialized oil-producing cells known as *elaiophores* have been discovered on the lip. Located in the callus, these cells secrete a slick of calorie-rich oil, a powerful attractant to oil-collecting bees since nutrients are difficult to acquire in tropical forests. *Rudolfiella aurantiaca* likely attracts pollinators in the same fashion, as do the flowers of several other genera (see *Gomesa*). The position of the elaiophores ensures that bees bump the underside of the column, fertilizing the flower. *Rudolfiella* was named after the German botanist and orchid fanatic Rudolf Schlechter (see *Dichaea*), whose brusque carriage and fierce determination rubbed more than a few contemporaries the wrong way. Renowned American orchidologist Oakes Ames once referred derisively to Schlechter's list of Central American species as "an undigested compilation which bristles with errors" (Ossenbach 2009: 142). Nevertheless, before his death in 1925 the German had amassed an enormous and widely respected collection of orchid specimens that was housed in Dahlem, near Berlin. Tragically, the collection was destroyed by Allied bombing raids during World War II.

Rudolfiella floribunda showing wings of lip
Andy Phillips

R. aurantiaca showing petals forming hood over column
Ecuagenera

Sarcoglottis

Description: Attractively patterned leaves, a fuzzy green inflorescence, and a curled, fleshy lip differentiate terrestrial *Sarcoglottis* orchids. **Flower:** Nearly tubular flowers emerge from lengthy, leaf-like bracts. Pale green to white, the medium-sized flowers (1–2" [2.5–5 cm]) are hairy on their exterior. The petals and dorsal sepal form a hood over the pointed column. Lateral sepals curl back, the tips often crossing, and are partly fused, enclosing the base of the lip. The outer edge of the thickened, tongue-shaped lip also curls down and back; *Sarcoglottis* is from the Greek for "fleshy tongue." **Plant:** Soft green leaves, spotted or striped with white, form a rosette near the ground. The fuzzy, upright inflorescence bears numerous flowers, often opening before the leaves emerge. **Similar:** See *Cyclopogon*, *Eltroplectris*, and *Pelexia*.

Distribution and Diversity: Occurring as terrestrials or rarely epiphtyes, nearly 50 species occupy varied conditions and a range of habitat types. The genus is found from Mexico and the Caribbean to Brazil and Argentina.

Ecology and History: The small, unpigmented flowers of *Sarcoglottis* are visited by euglossine bees that use the sepals as a landing pad. Nectar, rather than fragrance, provides the main attraction. Male and female *Euglossa* bees pollinate *S. fasciculata*, which provides a nectar reward in a spur hidden by the lateral sepals. To avoid hybridization between orchids served by the same pollinator, different species attach pollinia to distinct areas of insect bodies. One orchid might glue pollinia on a bee's head, another to its back: flowers of the first species cannot be fertilized by a bee carrying pollinia from the second. Nonetheless, the majority of insect pollinators receive pollinia on their upper side, on top of the thorax, abdomen, or head. In *Sarcoglottis*, however, pollinia are attached to the underside of the bee's head. They are glued to a mouthpart called the labrum, akin to our upper lip, that points down in normal orientation. This unique placement offers significant benefits. First, the pollinia are attached to an area difficult for the bees to clean, and thus are rarely removed. Second, the bees fold their lip before flying, moving the pollinia under the head and affording them protection during flight.

Sarcoglottis cf. *sceptrodes* fleshy lip and curled sepals
Andy Phillips

S. acaulis showing bracts beneath flowers
Ron Kaufmann

S. grandiflora showing fused sepal tips
Ron Kaufmann

Sarcoglottis foliage
Andy Phillips

Scaphosepalum

Description: *Scaphosepalum*, the Moustache Orchid, has unmistakable flowers with long tails projecting out to the sides or downward. **Flower:** These small, upside-down (non-resupinate) flowers usually are yellowish with dark purple markings, and take a variety of strange forms. Lateral sepals are nearly completely united, forming a bowl over the uppermost lip and column (the genus name means "bowl-sepal"), but with elongated, antennae-like tails extending outward or drooping down in most species. Each sepal bears a triangular fleshy pad on its interior surface near the tip. The thin dorsal sepal curls in front of the bowl. Petals are tiny and inconspicuous, as is the violin-shaped lip. **Plant:** Lacking pseudobulbs, diminutive plants grow as clumps of short stems with overlapping sheaths, each producing a single, thin leaf. Leaves narrow toward the base, joining short petioles. Long, sprawling inflorescences from the stem's middle produce 1 open flower after another, sometimes for several months. **Similar:** *Pleurothallis* and other relatives have inflorescences from the top or base of the stem, not the middle, and usually lack mustache-like tails.

Distribution and Diversity: Approximately 45 species grow epiphytically, occasionally terrestrially, in low- to high-altitude forests from Mexico to Bolivia.

Ecology and History: Like many of their relatives, *Scaphosepalum* flowers are pollinated by small flies attracted to the purplish, splotchy markings. Once alighted on the flower, some short-range guidance can be required to bring the tiny insect into contact with the pollinia. In *Scaphosepalum* and related groups (e.g., *Dracula, Masdevallia,* and *Pleurothallis*), the lip is hinged and lightly balanced, pushing the fly against the column when it passes the tipping point. As the captive struggles to escape, pollinia are attached. Recent DNA studies have revealed that the Andes, which rose swiftly over the past 10 million years, divided *Scaphosepalum* into two disjunct groups. One resides on the eastern foothills (e.g., *S. breve*) and is genetically isolated from the second (e.g., *S. microdactylum*), which ranges from the western slopes into Central America. The topography of the Andes has promoted the diversification of many South American orchids, much as oceanic barriers between the Galapagos Islands led to the rapid speciation of Darwin's finches.

Scaphosepalum antenniferum showing fleshy sepal pads
Valérie Léonard

S. grande
Andy Phillips

Scaphosepalum triceratops showing tiny, violin-shaped lip
Andy Phillips

S. decorum
Ron Kaufmann

S. fimbriatum, named for bristling (fimbriate) sepal tails
Ron Kaufmann

S. swertiifolium showing upward-curling dorsal sepal
Ron Kaufmann

Scaphyglottis

Description: *Scaphyglottis* are densely growing, dwarf plants best recognized by their thin, cylindrical pseudobulbs growing one atop the other in stacked chains. **Flower:** Sepals and petals are similar, slightly cupped rather than spreading. The lip is larger, with scalloped margins and a waterslide-like throat, giving the genus its name from the Greek words for "hollow" and "tongue." Flowers are small (to 1/2" [1.3 cm]) and pale yellow to greenish or purple, but bright reddish orange in hummingbird-pollinated species. **Plant:** Thin sheaths give rise to clustered stems composed of stacked pseudobulbs, themselves cylindrical to football-shaped. Short inflorescences emerge from the apex of a pseudobulb, which may be in the middle or at the end of a chain, and can produce large numbers of flowers. Narrow leaves (to 6" [15 cm] long), with a prominent central crease, arise in pairs or trios from atop the pseudobulbs.

Distribution and Diversity: Some 70 *Scaphyglottis* species occur in low- to middle-altitude humid forests, mostly as epiphytes in brightly sunlit treetops. They can be found from Mexico and the Caribbean to Brazil and Bolivia, with peak diversity in Costa Rica.

Ecology and History: While brightly colored flowers in this genus attract hummingbirds, the majority are pollinated by small wasps that arrive seeking nectar secreted on the lip, and depart with pollinia glued to their legs. Light-tolerant orchids like *Scaphyglottis* can sometimes survive outside of intact forest. One species, *S. livida*, colonizes coffee farms where legume trees (*Inga*) are planted to cast shade and replenish soil nutrients. The orchids establish on thin branches of these trees and successfully produce seed capsules, proof of the presence of pollinators. Such shade plantations offer more forest-like conditions than coffee grown in full sun, and provide habitat for migratory birds, anteaters, and many other tropical species. *Scaphyglottis livida* also has a long medicinal history in Mexico, where it is used to eliminate parasites, soothe stomach aches, relax muscles, and prevent miscarriages. Clinical trials showed that plant extracts produce pain relief and anti-inflammatory responses, an example of the nearly untapped potential of tropical biodiversity.

Scaphyglottis aurea, hummingbird-pollinated
Andy Phillips

S. fusiformis showing convex lip base
Andy Phillips

Stacked, cylindrical pseudobulbs of *S. modesta*
Ron Kaufmann

Scuticaria

Description: Attractively patterned flowers and long, whip-like leaves distinguish *Scuticaria*, which derives its name from the Greek word for "lash." **Flower:** Large (to 3" [7.5 cm]), showy blossoms are long-lived and strongly fragrant. Sepals and slightly smaller petals both are greenish yellow with reddish mottling or stripes. The 3-lobed lip is slightly cupped and often marked with radiating lines. Lateral lobes nearly encircle the column, while the fore lobe is notched and broadly flaring. A ridged, fleshy crest decorates the lip base, beneath a long column that ends in a downward-pointing cap. **Plant:** Long (to 48" [1.2 m]) leathery leaves are round in cross section and grooved along their length. Typically dangling, leaves arise singly, rarely in pairs, from pseudobulbs wrapped in tan sheaths; pseudobulbs are so narrowly cylindrical that they resemble stems. Inflorescences bear 1 to 3 flowers each. **Similar:** *Maxillaria* lip is oblong, jawbone-shaped. *Brassavola* flowers are white.

Distribution and Diversity: Inhabiting cool montane forest or humid lowlands, some 10 *Scuticaria* species grow epiphytically, terrestrially, or on rocks. They range from Venezuela and the Guianas to Peru and Brazil.

Ecology and History: Euglossine bees are believed to pollinate the attractive flowers of *Scuticaria*. Microscopic analysis of the lip surface reveals the presence of tiny hairs that form pseudopollen, a reward for the bees (see *Polystachya*, *Maxillaria*). Pseudopollen, common in species of *Maxillaria* visited by stingless meliponine bees, is rare in orchids pollinated by euglossines. We have mentioned already that orchids develop associations with fungi to assist germination (see *Ionopsis* and *Sobralia*) and nutrient delivery (see *Govenia* and *Mormodes*). *Scuticaria irwiniana* hosts a different fungus, *Fusarium oxysporum*, in its stems and leaves. This particular strain of *Fusarium* protects the orchid against harmful infections: laboratory studies confirm its antimicrobial and antifungal properties. *Fusarium* strains have extraordinarily diverse effects, however. One variety caused a worldwide banana blight that wiped out the Gros Michel cultivar worldwide; the Cavendish cultivar that we eat today (see *Galeandra*) was introduced in the 1990s as a resistant substitute. Unfortunately, new strains now attack Cavendish bananas, for which there currently is no replacement.

Scuticaria hadwenii
Andy Phillips

S. irwiniana showing lip's lobes and fleshy ridge (upper flower)
Ron Parsons

S. salesiana flower and whip-like leaves
Ron Kaufmann

Selenipedium

Description: The delightful, deeply pouched flowers of *Selenipedium* are unmistakable, borne on cane-like stems reaching 15 feet (4.6 m) in height. **Flower:** A highly modified lip, its edge broadly rolled inward, forms the slipper-like pouch. The rim's shape, like a crescent moon, gives rise to the genus name: Greek for "moon slipper." Narrow petals are dwarfed by the lip and the broad, somewhat hooded dorsal sepal. Lateral sepals are fused for most or all of their length, and held beneath the pouch. **Plant:** Many leaves emerge from two opposing sides of the tall, slender, occasionally branched stems. Leaves are pleated and lightly hairy. Inflorescences arise from on top of the stems, bearing multiple flowers; usually just one is open at a time. **Similar:** See *Phragmipedium.*

Distribution and Diversity: A half-dozen epiphytic and terrestrial species decorate grassy or forested low-elevation habitats. They are on display in Trinidad and from Panama to Peru and Brazil.

Ecology and History: The blossoms of *Selenipedium* are fantastically modified as trap flowers. They attract flies with a combination of narrow, sometimes twisting petals (see *Phragmipedium*) and disorderly splotches of bruised colors that call to mind rotting meat or vegetation. Flies seeking to lay their eggs on decaying carcasses are visually attracted to such cues (see *Bulbophyllum*). Having alighted on the lip, hapless insects are apt to slip off the rolled rim into the smooth pouch. The only exit is a series of stiff hairs at the rear of the sac, an escape ladder that guides the visitor directly under the short column where pollinia are smeared on its back or pressed into the stigmatic zone. Like other slipper orchids, *Selenipedium* has suffered at the hands of greedy fanatics, who call to mind collectors of past centuries in their willingness to strip whole forests bare. One of the tallest terrestrial orchids (and a vanilla substitute), *S. chica* is now exceedingly rare in Central America; currently, widespread deforestation is its principal threat. Happily, slipper orchid clubs have sprung up worldwide, working passionately to protect remaining habitat and to ensure that all commercially traded plants are produced only in greenhouses.

Selenipedium aequinoctiale, showing pouch-shaped lip with rolled rim
Ron Kaufmann

S. aequinoctiale plant; see flowers in top right
Ron Kaufmann

S. aequinoctiale, with developing seed capsule at left
Ron Kaufmann

Sievekingia

Description: *Sievekingia* ("see-ve-KING-ee-ah") plants resemble small versions of *Stanhopea*; however, the flowers are more compact and lack the latter's sweeping column and horned lip. **Flower:** Delicate flowers, despite the hanging inflorescence, are held horizontally with the lip uppermost (non-resupinate). Yellow to yellowish orange in color, the slightly concave sepals and smaller petals curl forward to partly enclose the column. Petals often are fringed, sometimes wildly so. A similar fringe decorates the outer lip margin of some species. The lip base forms a shallow pouch, in front of which is a fleshy hump (callus) sprouting upright teeth or vanes. **Plant:** Tightly clumped pseudobulbs are deeply furrowed, bearing papery bracts and a single leaf. Upright, pleated leaves with distinct veins are somewhat waxy and sparsely covered in short black hairs. The inflorescence dangles beneath the plant, producing up to 15 flowers. **Similar:** See *Paphinia*. *Stanhopea* flower faces downward, lip bears horns.

Distribution and Diversity: Dwelling epiphytically or on rocks, some 15 species favor lower elevations in humid, mountainous forest. The genus ranges from Costa Rica to Bolivia along the Andes; an isolated species (*S. jenmanii*) occurs north of the Amazon.

Ecology and History: *Sievekingia* species are pollinated by male euglossine bees (*Euglossa*) that land on the lip of these upside-down flowers, then crawl inverted into the blossom. In *S. jenmannii* and *S. suavis*, bees use the petals for their landing platform. The cupped shape of the flower blocks entry from the side, forcing a frontal assault and increasing the likelihood of successful pollination. Bees seek fragrance compounds that they use during courtship (see *Gongora* and *Stanhopea*), which are produced by glands in the depression at the lip's base. *Sievekingia* scents include limonene (oranges) and geraniol (geraniums). Access to these perfumes is impeded by the vaned callus that forces bees to climb the stout column. As they ascend, the bees' feet flick away the anther cap, exposing sticky pollinia that adhere to their legs. When visiting other flowers, these pollinia are pressed into the stigmatic area, achieving fertilization. Several species, including *S. fimbriata*, grow in ant nests, from which they may extract meager nutrients (see *Coryanthes*).

Sievekingia suavis
Andy Phillips

S. hirtzii showing pleated leaves
Andy Phillips

S. fimbriata with heavily fringed petals
Ron Parsons

Sobralia

Description: *Sobralia* orchids are notable for their large, showy flowers atop tall, slender, cane-like stems growing in clumps. **Flower:** The occasionally huge (1–7" [2.5–18 cm]) flowers are lavender, white, purple, or pink, less commonly yellow or orange. Sepals and petals are held open, the petals with wavy edges. A prominent lip partially encircles the column, trumpet-like, its outer edge typically wavy or frilly. The long, narrow column sometimes is winged near the tip. Often fragrant, flowers usually wilt in only 1 to 2 days. **Plant:** Strong stems, single or branched, often are clustered and range from 6 inches to 6 feet (15 cm to 1.8 m) tall; however, *S. dichotoma* may surpass 16 feet (5 m). Plants lack pseudobulbs, but produce fleshy roots. Leaves are papery or leathery and strongly veined, appearing ribbed. Flowers emerge from on top of the stem in nearly all species,1 to several per inflorescence. **Similar:** See *Elleanthus* and *Epistephium*.

Distribution and Diversity: More than 100 terrestrial, or rarely epiphytic, species decorate wet forest at all elevations, from Mexico to Peru, Bolivia, and Brazil.

Ecology and History: Most species are pollinated by female euglossine bees seeking nectar. Darwin described a greenhouse bee pollinating a *Sobralia*, stating that "the nectar of this Guatemala orchid seemed too powerful for our British bee, for it stretched out its legs and lay for a time as if dead on the labellum [lip], but afterwards recovered" (Darwin 1877: 92). When the bee backs out of the flower throat, it bumps the column and pollinia are affixed to its back. Variation in throat size curbs hybridization: large bees cannot enter, and small bees fail to brush the column. Environmental cues trigger synchronous flowering, with disparate species exhibiting staggered responses that further limit hybridization. *Sobralia* played a key role in orchid breeding. Typical orchid seeds contain virtually no stored energy. Development relies instead on symbiotic fungi that deliver nutrients from the environment. This phenomenon bedeviled early cultivators who could meet soaring demand only by selling wild-harvested orchids, fueling rampant overcollection. *Sobralia macrantha* produces uniquely large seeds, however, that germinate without fungus, and in 1853 it became one of the first sustainably bred commercial orchids.

Sobralia macrantha
Andy Phillips

S. macra showing trumpet-like, ruffled lip
Andy Phillips

Sobralia fimbriata
Andy Phillips

S. atropubescens showing winged column
Andy Phillips

S. sessilis
Andy Phillips

S. pulcherrima showing stem and pleated leaves
Ron Kaufmann

Stanhopea

Description: *Stanhopea* flowers are distinguished by twin fleshy horns projecting from the middle of the lip, earning the nickname *El Torito*, or "little bull." **Flower:** Large, flaring sepals curl back at the tips, as do smaller petals. Both are usually creamy or yellow, with purplish mottling. The long, thin, and arching column bears wings near the tip. The waxy, downward-hanging lip is a complex, 3-part structure. The lowest region is glisteningly smooth and usually pointed; the narrow central area bears twin horns, ranging in size from goat to longhorn steer; the base is concave, resembling a wooden clog. Most species have large (to 8" [20 cm]), strongly perfumed flowers. **Plant:** Pseudobulbs are tall (2–3" [5–7.5 cm]), fluted, and wrapped by fibrous sheaths when young. From atop each pseudobulb arises a large, stiffly pleated leaf with a distinct petiole. The inflorescence of 2 to 12 flowers or more hangs below the pseudobulb base. **Similar:** See *Acineta*, *Gongora*, and *Sievekingia*.

Distribution and Diversity: More than 60 species occur as epiphytes and a few terrestrials in low- to middle-altitude humid forests. They are found from Mexico and Trinidad to Bolivia and Brazil.

Ecology and History: Flowers advertise their brief availability with strong scents, beckoning male euglossine bees that collect fragrance blends exuded near the lip base. Some claim these odors intoxicate the bees, hampering their flying ability. Others dispute the drunken bee theory, but bees do routinely lose their grip and fall down the slippery lip, herded between the horns. Plummeting, they brush the column and pick up, or deposit, pollinia. One evolutionary problem with fragrance rewards is that pollinators return repeatedly to the same flower, which risks self-fertilization. *Stanhopea* avoids this through sequential male and female floral phases. Once oversized pollinia are removed, the undersized stigmatic area enlarges; meanwhile, drying pollinia shrivel until a fit (in a second flower) can be achieved. Flowers wilt in 2 to 4 days, saving energy. *Stanhopea anfracta*, an inhabitant of stormy mountainsides, remains open a few days longer, however, awaiting the rare sunny day when fragrance-collecting bees are active. These scents are harvested not only by bees: Mexicans gather *S. tigrina* to perfume corn tortillas, while Ecuadorians use eucalyptus-scented *S. anfracta* to treat coughs.

Stanhopea gibbosa showing lip horns
Andy Phillips

S. tigrina hanging inflorescence and foliage
Ron Kaufmann

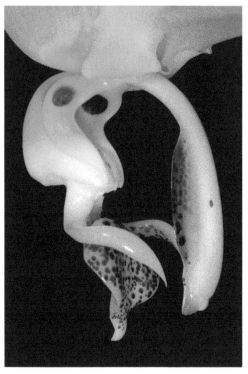

Stanhopea embreei close-up of lip (left) and column
Andy Phillips

S. wardii
Andy Phillips

S. insignis showing column wings
Andy Phillips

Stanhopea furrowed pseudobulbs and fibrous sheaths
Ron Kaufmann

Stelis

Description: An enormously diverse genus, *Stelis* can be recognized by the combination of small, flat, usually triangular flowers, and leaf stems with upturned trumpet-like sheaths. **Flower:** Sepals are similar, held open, and fused for part of their length, giving a strongly 3-sided appearance (even buds are triangular). Somewhat rounded at the tips, sepals may be covered in bumps or hairs. Flowers vary in color, but typically are thin and translucent. Petals, lip, and column are exceptionally short. **Plant:** Fleshy or leathery leaves emerge from erect stems encircled by a repeating series of trombone-shaped sheaths. Pseudobulbs are absent. Bearing multiple flowers, the slender inflorescence emerges from the leaf base. **Similar:** See *Lepanthes* and *Pleurothallis*.

Distribution and Diversity: More than 800 species of *Stelis* occupy nearly every habitat type and elevation. The genus is found in Florida and the Caribbean, Mexico to Brazil and Bolivia. Epiphytic and terrestrial, they largely prefer cool, moist forests, reaching their peak diversity in the Andes.

Ecology and History: The delicate flowers of *Stelis* show diverse adaptations to pollination by flies. Most flowers produce nectar, often visible as clear droplets, to reward their fly visitors, and many emit a mild, attractive odor. Numerous species have light-sensitive flowers that require direct sunlight to open; at night, the flowers close completely, presumably to save valued fragrance compounds for the next day. Others release scent only during periods of high humidity, closing during the hot, dry hours of midday. Still other species, particularly in Mexico, are open only at night. In all cases the flowers are programmed to release fragrance when their pollinators are active. If pollinators are absent, some species are capable of self-fertilization. *Stelis cleistogama* can fertilize itself even while the flower is still in bud, a process known as *cleistogamy*. Cues other than scent may be employed, for example, *S. villosa* captures the attention of flies with icicle-like waxy strings that hang from its sepals, swaying in the faintest breeze. Other flowers bear radiating lines that steer visitors toward the bloom's center. In *S. argentata* these lines are decorated with reflective calcium oxalate crystals (see *Lepanthes*), a shiny trail to which flies respond strongly.

Stelis inflorescence in habitat
Joe Meisel

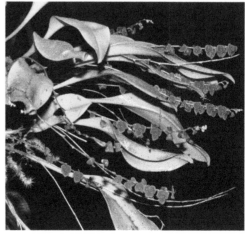

Stelis cf. *purpurea* leaves and flowers
Ron Kaufmann

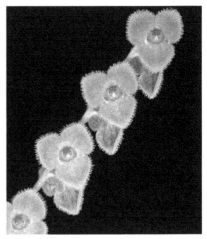

Stelis ciliolata showing tiny lip and petals
Andy Phillips

S. cf. *argentata* with reflective crystals on
sepals
Andy Phillips

S. villosa attracts fly pollinators with dangling threads
Andy Phillips

Stelis inflorescence showing translucent sepals
Ron Kaufmann

S. pilosa showing range of forms in genus
Andy Phillips

Stenia

Description: Curious-looking flowers with lips pinched into a tunnel or bucket, and short, thin leaves characterize *Stenia*. **Flower:** Large (to 3" [7.5 cm]) flowers are white to pale greenish yellow, often with purple spotting. Sepals and petals are similar, lateral sepals occasionally larger and more pointed. The lip varies from concave or tunnel-like to deeply pouched, with a pinched or rolled frontal opening. Variably shaped side lobes of the lip are heavily spotted with red. At the base of the lip repose a few toothed keels. **Plant:** Clustered stems are sheathed by leaf bases; leaves are arranged in a loose fan. The short, curving inflorescence arises from the sheaths, hoisting a single flower. **Similar:** See *Chondrorhyncha*, *Cochleanthes*, and *Huntleya*.

Distribution and Diversity: Middle- to upper-elevation wet forests are favored by more than 20 epiphytic species. They occur from Trinidad and Venezuela to Brazil and Bolivia

Ecology and History: The pollinators of *Stenia*'s peculiar flowers are unknown, but euglossine bees are strongly suspected. This mystery underscores how much remains to be revealed about orchids: new discoveries unquestionably await any botanists who dedicate themselves to this genus. *Stenia dodsoniana* pays homage to modern orchid expert Calaway "Cal" Dodson (born 1928), author of the breathtaking *Native Ecuadorian Orchids*. In the 1970s he established the Rio Palenque Science Center in the biodiverse lowlands of western Ecuador. An example of small-scale conservation, Rio Palenque shows how much can be protected by the efforts of just one person: its modest forest (250 acres [100 ha]) is home to more than 350 species of birds and more than 1200 species of plants, one-quarter of which occur nowhere else on Earth. Sadly, Rio Palenque also illustrates the gloomy consequences of a landscape dominated by commercial agriculture. Formerly connected to larger, species-rich forests, the reserve became surrounded by sprawling plantations of bananas, and later, African oil palms now in high demand by snack food and biofuel industries. With all links to other woodlands severed, the reserve's isolated plants and animals are suffering: more than a dozen bird species have disappeared, and declines in orchids dependent on specialist pollinators are sure to follow, a pattern confirmed in other fragmented tropical forests.

Stenia lillianae
Andy Phillips

S. lillianae from side
Andy Phillips

S. pallida showing pinched lip and rolled edges
Ron Parsons

Sudamerlycaste

Description: Pleated leaves and large, nodding flowers with frilly lips distinguish *Sudamerlycaste*. **Flower:** Tilting downward and slightly bell-shaped, the large (to 4" [10 cm] across), waxy flowers are green, yellowish orange, or beige. Lateral sepals are sickle-shaped, while smaller petals spread outward or hood the slender column. The lip's outer portion is densely fringed, the inner region bears 2 upright wings; lip color ranges from green to orange or red. **Plant:** Pseudobulbs are tall and tapering, or egg-shaped, each bearing 2 to 4 broad, pleated leaves. A thick inflorescence with numerous leafy bracts emerges from the pseudobulb base, typically yielding a single flower. **Similar:** See *Anguloa* and *Lycaste*.

Distribution and Diversity: Some 25 epiphytic and terrestrial species occupy varied wet habitats. The genus occurs in the Caribbean and from Venezuela to Bolivia and Brazil.

Ecology and History: The dull-colored flowers of *Sudamerlycaste* are strongly fragrant at dusk and night. Some produce nectar at the base of the lip, dripping down the sepals and petals in *S. locusta*. These traits suggest moth pollination, and moths have been observed visiting the flowers but not removing pollinia. Nocturnal bees, however, are known to accomplish fertilization. Nectar production in *S. cinnabarina* peaks by day, however, hinting at a different purpose: it may attract ants to protect flowers from herbivores. *Sudamerlycaste reichenbachii* commemorates a troubled legend in orchid history, German taxonomist Heinrich Gustav Reichenbach (1823–1889). His superb series of orchid paintings *Reichenbachia* cemented his reputation as the most prominent orchid expert after Lindley (see *Barkeria*). He was consumed by orchids, as his obituary summarized: "To him meals and clothes were necessary evils, but his herbarium was a prime necessity of existence" (Reinikka 1995: 216). He amassed an enormous collection of orchid specimens, notes, and sketches, but restricted access to his famed herbarium with notorious stinginess. Upon death he willed his collection to the Vienna Imperial Museum, and in a fit of petty jealousy demanded that Vienna seal the herbarium for 25 years, an "obscurantist interregnum" that denied fellow scientists a trove of information. Cantankerous by nature, he dismissed the tribute of orchids named for him, derisively commenting, "I cannot eat the honor."

Sudamerlycaste heynderycxii
Ron Kaufmann

S. reichenbachii showing lip wings
Ron Parsons

S. ciliata pseudobulb (left), leaves
Ron Kaufmann

Telipogon

Description: Striped petals provide a backdrop for the remarkable fly mimicry of *Telipogon*. **Flower:** The stout, stubby column sprouts fringes or clumped bristles (the genus name derives from the Greek for "bearded tip"), giving the appearance of a lifelike fly. Medium-sized (2" [5 cm]) to nearly microscopic (1/8" [0.3 cm]) flowers are spreading and painted with a variety of bright colors. Petals, larger than the sepals, typically are marked with radiating purplish lines. The broad lip is uppermost (non-resupinate) in nearly all flowers. Pollinia are attached to a boathook-like structure visibly protruding from the column. **Plant:** Small plants often are dwarfed by their flowers. Pseudobulbs when present are inconspicuous, covered by sheaths. Stems branch and scramble, bearing a few short, fleshy leaves that emerge from opposite sides of the stem or from its base; leaves are deciduous at flowering in some species. **Similar:** *Trichoceros* has obvious pseudobulbs, a longer, wiry inflorescence, and a densely spiky column.

Distribution and Diversity: More than 100 species thrive in medium- to high-elevation sites as epiphytes or terrestrials on mossy ground. They range from Mexico and the Caribbean to Bolivia; peak diversity is in the Colombian Andes.

Ecology and History: These orchids, and a few others, are called *La Mosca* (the fly) for their uncanny imitation of bristly tachinid flies that serve as pollinators. Tachinids are large bodied, have excellent vision, and exhibit complex reproductive behaviors. Male tachinids patrol a territory from which they aggressively evict other males. Receptive females alight within the territory and display their readiness to mate, whereupon the male zooms down and copulates with her. Confused males grappling amorously with *Telipogon* flowers are snared by the boathook, unwittingly pulling the sticky pollinia onto their belly. This strategy, known as *pseudocopulation*, is used by other orchids to lure gnats (see *Lepanthes*), flies (*Trichoceros*), and wasps. *Telipogon* relies on visual deceit rather than sex pheromones (see *Restrepia*), and males may simply be attacking a supposed rival (pseudoantagonism). The radiating petal stripes also help catch a fly's eye: insect compound eyes are sensitive to alternating light and dark bars that produce a flickering effect as the bug speeds past.

Telipogon ionopogon showing boathook pollinia
Ron Parsons

T. microglossus
Ron Kaufmann

T. hausmannianus
Ron Parsons

Trichocentrum

Description: Eye-catching, large flowers (to 3" [7.5 cm]), often with a conical spur extending from the lip, and thick or cylindrical leaves characterize *Trichocentrum*. **Flower:** Sepals and petals are similar in shape and color: a base of cream, yellow, or green, often splotched or suffused with brownish red. The 3-lobed or violin-shaped lip is the showiest element, painted with white, yellow, or purple. The spur can be a short bulge or a long thread (*Trichocentrum* is from the Greek for "hair-like spur"). Two tattered wings emerge from the short column's apex. Keel-like ridges or massive teeth decorate the lip base. **Plant:** Tiny pseudobulbs are obscured by sheathing leaf bases. Leaves are fleshy and leathery, folded at the base like a mule's ear or cylindrical with a narrow groove. Some are quite purple underneath. Hanging or arching inflorescences emerge from the bases of pseudobulbs, bearing few to many flowers. **Similar:** See *Oncidium*.

Distribution and Diversity: Favoring shady low- to middle-altitude forests, more than 70 species of *Trichocentrum* are primarily epiphytic, plus a few terrestrials. The genus ranges from Florida, the Caribbean, and Mexico to Bolivia, Paraguay, Argentina, and Brazil.

Ecology and History: Viewing *Trichocentrum* is an example of when not to trust your eyes. The genus likely evolved in dry, seasonal climates: leaves are succulent and hard skinned—common adaptations to drought—and deciduous during extended dry periods. Today, however, only a few species still inhabit such regions. The flowers often have a long spur, but it produces no nectar and does not attract long-tongued butterflies or moths as one might expect. Instead, the lip surface furnishes oils that attract euglossine bees; steered headfirst into the flower by the raised keels, the bees receive pollinia glued to their face. Oils are produced inside microscopic hairs that can be broken open only by particular bee species, or in saddle-like glands that smear oil on the lip keels. Most flowers are fragrant, but only during the day, to entice the bees which themselves are active only in bright sunlight. One widespread species, *T. cebolleta*, contains hallucinogenic alkaloids used by the Tarahumara people, Mexico's famed long-distance runners, in religious rituals.

Trichocentrum lanceanum showing 3-lobed lip
Ron Kaufmann

T. pulchrum showing column wings
Ron Parsons

T. stramineum hanging inflorescence
Andy Phillips

Trichoceros

Description: Flowers of several orchids mimic flies, but astonishing *Trichoceros* blossoms look more like a fly imitating a flower. **Flower:** A stout column is heavily bristled (the genus name refers to the Greek words for "hair horn"). The fleshy lip is large and arresting, with a broad central lobe and 2 narrow side lobes: the fly's outstretched wings. Similar petals and sepals are spreading; both are reddish or yellow, with radiating wine-colored stripes. **Plant:** Small, plump pseudobulbs are widely spaced, partly obscured by grayish sheaths producing fleshy leaves. In a few species the leaves are pleated rather than leathery. An occasionally branching inflorescence bears successively opening flowers. **Similar:** See *Telipogon*.

Distribution and Diversity: Some 10 species of *Trichoceros* grow epiphytically, or in a jumble on the ground, in high-elevation forests. Their range extends only from Colombia to Bolivia.

Ecology and History: *Trichoceros* employs visual mimicry and pseudocopulation to achieve pollination by tachinid flies (see *Telipogon*). In *T. antennifer* the simulation of a receptive female is so intricate that the shiny stigmatic area of the flower imitates her genital orifice: in nature such females signal sexual readiness by pulsing this opening. As the flower waves back and forth on its slender stem, this shiny area winks on and off, attracting a male tachinid. During his copulatory ministrations he contacts the pollinia, which are then glued to his abdomen. The stalk of the pollinia initially is erect, but curls forward after removal from the flower (this action is reversed in *Telipogon*). Only pollinia with fully curved stalks can fertilize another flower, a trick to prevent self-pollination. Although flies typically do not make good pollinators (see *Octomeria*), many high-elevation orchids rely on them because their favored pollinators, bees, decline in abundance with increasing altitude. In fact, flies pollinate between 15 and 25% of all orchids. Tachinids are unusually strong fliers, but do not seek nectar and are not normally attracted to flowers. Rather they are driven to reproduce, often laying eggs on caterpillars that later are consumed inside-out by their maggots. *Trichoceros* has evolved a powerful sexual lure for these flies, inducing them to move swiftly and efficiently between flowers.

Trichoceros antennifer
Ron Parsons

T. cf. *onaensis* showing 3-lobed lip
Andy Phillips

T. muralis
Ron Parsons

Trichopilia

Description: *Trichopilia* flowers are sometimes called Corkscrew Orchids for their wavy, twisting sepals and petals. **Flower:** Flowers are large (to 6" [15 cm]), white to greenish yellow, sometimes overlaid with brown or red. Narrow sepals and petals are similar, occasionally twirled, and symmetrically arranged in a single plane. The showy, funnel-shaped lip, typically white with bright yellow, pink, or red markings, projects forward; its outer margin is often wavy or pointed. A slender column is hidden by the rolled lip, but when viewed it displays a distinctive fringe crowning its tip (*Trichopilia* combines the Greek words for "hairy cap"). **Plant:** Strongly flattened, oval or linear pseudobulbs are densely clustered. Each yields 1 leaf from the top, and a hanging inflorescence from the base. Leaves are broad and leathery, sometimes quite long (to 12" [30 cm]), but species are vegetatively variable. **Similar:** *Brassavola* leaves are cylindrical. See *Cischweinfia*.

Distribution and Diversity: Epiphytes, and uncommon terrestrials, in middle- to high-altitude wet forests comprise approximately 40 species. They can be enjoyed from Mexico and the Caribbean to Brazil and Bolivia.

Ecology and History: Several *Trichopilia* species (e.g., *T. maculata*, *T. rostrata*) are pollinated by euglossine bees (various *Euglossa* species), presumably visiting the pleasantly perfumed flowers to collect scents. While orchids employ a variety of techniques to ensure they are not fertilized by pollinia of another species, sometimes these mechanisms fail and natural hybrids are formed. One example occurs in Costa Rica, where *T. marginata* and *T. suavis* are cross-pollinated, forming a hybrid known as *Trichopilia* × *crispa* (the "×" denoting a natural hybrid): its flowers are intermediate between the two parent species. While rare in nature, artificial hybrids are relentlessly generated by hobbyists and commercial nurseries searching for appealing and marketable combinations of traits from different species. One parent may provide the overall shape, another supplies color, and a third the fragrance. Amazingly, some commercial hybrids involve crossing 8 distinct genera. Even more incredible: the total count of registered hybrids vastly exceeds the number of true orchid species on the planet (see *Barkeria* for a reaction).

Trichopilia tortilis
Andy Phillips

T. crispa showing fringed column tip and flattened pseudobulbs
Ron Kaufmann

Trichosalpinx

Description: *Trichosalpinx* orchids are distinguished by slender stems bearing successive trumpet-shaped sheaths and tiny flowers with blunt, unlobed petals. **Flower:** Relatively simple, small (to 1" [2.5 cm] across) flowers are purplish red to yellowish green. Sepals are much larger than the petals, occasionally fringed and often greatly elongated. Petals are simple and unlobed. The tiny lip's outer edge often is hairy. **Plant:** Tufted plants with erect stems bear leaves that occasionally are thick and fleshy; normally green, foliage of some species is attractively wine colored. The genus name, drawn from the Greek words for "hairy trumpets," is fitting: stems are sheathed with overlapping, ribbed, and bristly funnels shaped like upturned trumpets. Inflorescences emerge from the funnels, and yield 1 to 12 flowers at a time; as the blossoms drop off, they leave behind a fishbone-like pattern of flower stalks. **Similar:** A few related genera share the trumpet-shaped sheaths: see *Lepanthes*.

Distribution and Diversity: More than 100 species grow epiphytically, terrestrially, or on rocks, in low- to high-elevation sites from Mexico to Bolivia and Brazil.

Ecology and History: Many species are pleasantly fragrant, including *T. robledorum* (aka *Tubella robledorum*), which smells of citrus. Flowers of related genera are largely fly-pollinated, and the shape and color of *Trichosalpinx* blooms point to adaptation to flies, but the identity of the true pollinators remains a mystery. In tiny forest remnants in Mexico, the proportion of *T. blaisdelli* flowers successfully producing fruits is extremely low, suggesting that their unknown pollinators suffer declines in fragmented landscapes. Evidence of *Trichosalpinx* in tropical America first emerges in a crude woodcut dating to 1588, which clearly depicts the distinctive funnel sheaths. Numerous species in this large genus honor explorers and botanists, perhaps none more accomplished than Alwyn Gentry (1945–1993), memorialized by *T. gentryi* (aka *Tubella gentryi*). Gentry was one of the greatest modern tropical botanists, with an encyclopedic knowledge of South America's flora. He collected specimens relentlessly, was a tireless supporter of conservation, and published the definitive *Guide to Woody Plants of Northwest South America*. Tragically, he perished in a plane crash along with Ted Parker, South America's finest birder, while conducting conservation surveys in western Ecuador.

Trichosalpinx patula showing petals
Andy Phillips

T. robledorum
Andy Phillips

"Hairy trumpets" of *T. escobarii*
Joe Meisel

Trigonidium

Description: The flowers of *Trigonidium* recall a flamboyant Mardi Gras mask: their flaring, triangular shape focusing attention on a pair of eyes peering out from the center. **Flower:** Large sepals are yellowish green to brown with dark markings and are partly fused, creating a deep tube. Short petals are nested within the triangular opening (*Trigonidium* references the Greek word for "three-cornered"). Pointed petal tips are painted deep purple, or in some species, a reflective cyan. The pale, bulbous tip of the largely concealed lip sometimes is visible as the mask's nose. **Plant:** Short, ridged pseudobulbs, sometimes densely clustered, bear 1 to 3 leaves. Leathery leaves often are bent, drooping at the middle. One or more wiry inflorescences arise from the pseudobulb base, each supporting a single flower. **Similar:** *Maxillaria* flowers lack the strikingly triangular shape (although DNA evidence suggests the genera may be combined).

Distribution and Diversity: About a dozen species parade as epiphytes and terrestrials in low- to middle-altitude wet forests or scrub. The genus can be found from Mexico to Peru and Brazil.

Ecology and History: Plants of *Trigonidium* seduce small bees into mating with their flowers, the first documented case of pseudocopulation by bees. Flowers lure male meliponine bees with a combination of visual cues (the "eyes") and complex fragrances. Most flowers exude a citrus-like odor at midday, a unique blend of scent components targeting a single bee species. Meliponines, or stingless honeybees, are social insects that produce large numbers of reproductive males just a few times each year. To ensure that flowers will be open when males emerge, a single *Trigonidium* plant can produce flowers for up to 7 months, although individual blooms last only 4 to 10 days. The bees alight on the flower and attempt to copulate with the sepals or petal tips. Less sure-footed bees occasionally slide down the waxy interior of the tunnel, where the flexible lip pins him gently against the column. While struggling to escape, pollinia are glued to his back. The pollinia, swollen when fresh, dry quickly and shrivel enough to fit the narrow stigmatic cavity of a second flower, reducing the risk of self-pollination.

Trigonidium obtusum showing pale lip
Andy Phillips

T. acuminatum showing reflective blue petal tips
Andy Phillips

T. grande from Ecuador
Andy Phillips

Trisetella

Description: The dwarf plants of *Trisetella* exhibit flowers usually shaped like a purple tongue with twin yellow tails. **Flower:** Thinly textured blossoms are dull yellow, orange, or white, often with substantial maroon markings. The somewhat broad dorsal sepal is joined at its base to the lateral sepals, forming a shallow cup; lateral sepals are completely fused (except in *T. pantex* and *T. hoeijeri*) into an oblong tongue. All 3 sepals display trailing tips in most species, earning *Trisetella* its name meaning "three little bristles." Huddled around the hooded column are tiny, thin petals and an arrowhead-shaped lip. **Plant:** Short, clumped stems bear narrow, leathery leaves. Inflorescences arise from the stem apex, near the leaf bases, and produce numerous flowers, although only 1 is open at a time. **Similar:** *Masdevallia* and *Dracula* flowers are clearly 3-parted. *Scaphosepalum* flowers are non-resupinate.

Distribution and Diversity: Middle-elevation wet forests are punctuated by more than 20 species of epiphytes and terrestrials, from Costa Rica to Peru and Brazil.

Ecology and History: *Trisetella* flowers, with their abundant purple markings, longitudinal stripes, and elongated tails, display all the hallmarks of fly pollination. But what separates them from other fly-pollinated orchids such as *Masdevallia*? This question lies at the heart of taxonomy, the science of organizing species into groups based on shared characteristics. For centuries, species were classified using visible, structural characteristics known collectively as *morphology*. The traditional orchid classification system was based on morphology and was devised largely by Robert Dressler (born 1945), for whom *T. dressleri* is named (see *Dresslerella*). Orchids may visually resemble one another, however, due to similar evolutionary pressures rather than shared ancestry. Hummingbird-pollinated flowers, for example, are reddish orange and tubular, but occur in many genera. Molecular taxonomy examines DNA sequences that do not determine morphology, and thus are not influenced by evolution. These sequences accumulate random mutations over time, like modifications to a spoken message in the childhood game "Telephone." The sum of such changes is an unbiased measure of the relatedness of any two orchids. *Trisetella* species show highly similar sequences, but as a group differ from *Masdevallia* species: here genetic and morphological methods are in agreement.

T. strumosa showing tiny lip
Andy Phillips

Trisetella dressleri
Ron Parsons

T. hoeijeri, a species with unfused sepals
Ron Kaufmann

Vanilla

Description: *Vanilla* plants are easily recognized by their climbing, vine-like appearance and spectacular but short-lived flowers. **Flower:** Spreading sepals and petals of these large flowers (to 4" [10 cm] across) are pale yellowish green to white. The forward-jutting lip, usually marked with orange or yellow, occasionally red to purple, has a ruffled outer edge and a sheath-like base that wraps the column. The long column is hairy underneath, with a large flap of tissue (the rostellum) covering the stigmatic area. **Plant:** A very elongated stem (to 40' [12 m] or more) twines up tree trunks and branches, clinging with aerial roots. Leaves can be large (to 9" [23 cm] and longer), shiny, and leathery, or so small the plant resembles a leafless vine. Inflorescences emerge from the leaf axils, bearing only a few flowers that wilt within hours of opening.

Distribution and Diversity: More than 100 species occur terrestrially, or rarely as epiphytes, throughout the world's low- and middle-elevation wet tropical forests.

Ecology and History: Vanilla is the most important commercial product obtained from orchids. The flavoring, composed of vanillin plus more than 150 aromatic compounds, is produced through labor-intensive processing of vanilla "beans" (actually seed capsules: the tiny black flecks in natural vanilla ice cream are orchid seeds). Commercial vanilla comes primarily from *V. planifolia*, which in the wild is pollinated by bees. Large-scale production was hampered until a slave named Edmond Albius discovered that lifting the rostellum with a sliver of bamboo permitted hand-pollination. Since his epiphany, most operations moved to Madagascar and Indonesia, together responsible for 80% of world production. With prices occasionally exceeding US$1,000 per pound, vanilla is the second-most expensive spice after saffron. The Aztecs mixed vanilla with cacao to make a sumptuous drink they called *chocolatl*. Spanish conquerors brought it home, and soon hot chocolate became popular in cafés across Europe. Thomas Jefferson shipped the first cured capsules to the United States and published his recipe for ice cream (1 quart heavy cream, 1 vanilla bean, 6 egg yolks, and 1 cup sugar). Vanilla, taken alone or in chocolatl, is reputed to have medicinal and aphrodisiacal properties: the Aztec emperor Moctezuma drank 50 pitchers per day, particularly before visiting his wives.

Vanilla aphylla showing rolled, ruffled lip
Andy Phillips

V. imperialis
Andy Phillips

V. planifolia, source of commercial vanilla
Andy Phillips

Warczewiczella

Description: Fan-shaped plants of *Warczewiczella* ("VAR-sheh-vitch-EH-la") display large flowers (to 4" [10 cm] across), the lip wrapping the column and adorned with a raised, grooved shelf. **Flower:** Sepals and petals are pale green or cream colored; sepals usually are swept sharply backward, partly rolled at the bottom, while petal tips are slightly curled. The flared, pale lip is decorated with yellow or violet blotches, or purple radiating lines. Its broad sides curve gracefully around the blunt-winged column. A distinctive platform, the callus, occupies the lip's base: its edges overhanging like a kitchen table, its surface furrowed into parallel, rounded ridges. Most species are strongly fragrant. **Plant:** Soft leaves with a central crease, arranged in fans, emerge from multiple sheaths that wrap the stem; pseudobulbs are absent. Stout inflorescences of a single flower arise from the lower sheaths. **Similar:** See *Chondrorhyncha* and *Cochleanthes*.

Distribution and Diversity: Approximately 10 species (many formerly in *Cochleanthes*) dwell epiphytically, or rarely terrestrially, in low- to middle-altitude humid forests. They range from Honduras to Colombia, Bolivia, and Brazil.

Ecology and History: Most *Warczewiczella* flowers are deceit-pollinated by euglossine bees. Attracted by potent fragrances (e.g., *W. amazonica*), the bee squeezes deep into the flower, where notches in the lip permit it to probe the nearly tubular sepal bases in search of nectar. Finding none, it backs out and brushes the column while navigating over the ample callus. Many mimic the nectar-rich flowers of vines, further deceiving bees into expecting a sugary reward. *Warczewiczella marginata* imitates flowers of a common bean vine (*Clitoria*, the Butterfly Pea). Male and female bees (*Eulaema meriana*) insert long tongues into the flower, but come away empty-mouthed; most pollinators are inexperienced youngsters, suggesting that older bees have learned the hoax. The genus honors Polish botanist and collector Josef Warszewicz (1812–1866); virtually all the many plants named for him (including this genus) variously misspell his cumbersome name. Fond of facial hair, he once was described as "a great traveller and one of the first botanists in the world. His name is Warscewicz [sic] . . . and wears a beard, in fact, is all hair, from his nose downwards!" (Ossenbach 2009: 85).

Warczewiczella discolor showing distinctive callus
Andy Phillips

W. amazonica
Andy Phillips

W. wailesiana showing column wings
Andy Phillips

Warrea

Description: Attractive, cupped flowers and jointed pseudobulbs characterize terrestrial *Warrea*. **Flower:** Sepals and petals are similar, concave and curling forward into a partial bell. The somewhat fleshy, medium-sized (1–3" [2.5–7.5 cm]), mostly nodding flowers are long-lived and fragrant. Colors range from white to yellowish orange, beneath a blush or stripes of magenta. The rim of the broad lip is rolled downward, while twin lobes at its base clasp an elongated column. Beneath the column a prominent and irregularly shaped mound sprouts from the lip. **Plant:** Clustered pseudobulbs are marked with transverse seams, and often are wrapped by papery sheaths. Several broad, thin, and pleated leaves arise from atop the pseudobulb, their bases narrowing to stiff stalks. **Similar:** *Xylobium* pseudobulb lacks seams. *Catasetum* and *Cyrtopodium* pseudobulbs are cigar-shaped.

Distribution and Diversity: Just 4 terrestrial species decorate low-elevation hillside forest and scrub, from southern Mexico to Brazil and Argentina.

Ecology and History: The perfumed flowers of *Warrea* are presumed to be pollinated by male euglossine bees (see *Gongora*). The position of the column and the mound-like keel beneath it resemble other orchids, which force bees to scrape their backs against the column's reproductive organs. The broad leaves, characteristic of many terrestrial plants, are an adaptation to gathering light on the dim forest floor. Although the genus name refers to little-known English collector Frederick Warre, one species, *W. hookeriana*, commemorates a true giant of British science: Joseph Dalton Hooker (1817–1911). Hooker, one of the century's finest botanists, was a tenacious explorer and Charles Darwin's greatest friend. The two carried on a prodigious and stimulating correspondence, while Hooker supplied countless orchids for Darwin's research (see *Catasetum* and *Zootrophion*). Despite his own brilliance, Hooker referred to himself jokingly as "a dull dog, a very dull dog," in comparison to Darwin's genius. Later he became an ardent champion of evolution, defending (alongside Thomas Huxley) his friend's theory in a famous verbal battle with Bishop Wilberforce: "I proceeded to demonstrate . . . that he [Wilberforce] could never have read your book and that he was absolutely ignorant of the rudiments of Botanical Science" (Cockerell 1919: 529).

Warrea warreana
Andy Phillips

W. warreana flower showing cupped sepals
Juan Carlos Uribe

Xylobium

Description: The tufted plants of *Xylobium* are characterized by long-stemmed, pleated leaves and clusters of pale flowers with distinctive warty lips. **Flower:** Lightly fragrant flowers are pale yellow to white, occasionally with purple speckling. Sepals and petals are similar, narrow and pointed. The downward-curving lip may match the sepals' color, but commonly is a rich purple. Often the outer edge is curled up, like a person who rolls their tongue. The lip's surface usually is covered in warty bumps, coalescing into a knobby hump at the base. The column is short and slightly curved. **Plant:** Pseudobulbs are round in cross section, from pear- to pencil-shaped, their bases wrapped by dry, opposed sheaths. One to 4 leaves arise from the pseudobulb apex; a short, arching or upright inflorescence emerges from the base. Leaves are leathery and pleated, visibly veined on the underside. **Similar:** See *Lycaste* and *Warrea*.

Distribution and Diversity: Roughly 30 species grow epiphytically, or rarely terrestrially, in low- to middle-altitude wet forests. They can be found from Mexico and the Caribbean to Bolivia and Brazil.

Ecology and History: The scented flowers of *Xylobium* provide no nectar, but several species (*X. ornatum*, *X. squalens*, and *X. variegatum*) attract social, stingless bees (*Trigona*) as pollinators. The bees land on the lip, scrabble at its surface, and depart by backing out. As they exit, a projecting shelf atop their back, the scutellum, bumps the tip of the column and 4 pollinia are glued to its edge. Pollination by social meliponine bees such as *Trigona* presents an interesting challenge, compared with the solitary euglossines that commonly service orchids. Social bees live in a nest with thousands of closely related sisters, and have a habit of routinely grooming themselves and each other to remove parasites and dangerous fungal spores that could infect the colony. The scutellar placement of *Xylobium* pollinia seems adapted to prevent a bee from grooming them off of its own back. Furthermore, the 4 pollinia are arranged so the upper pair covers the lower pair, the quartet forming a compact bundle that is difficult for other bees to remove.

Xylobium pallidiflorum showing cylindrical pseudobulb and curled lip
Andy Phillips

X. leontoglossum
Ron Kaufmann

X. squalens showing warty, purple lip
Andy Phillips

Zootrophion

Description: The genus *Zootrophion*, its name meaning "animal menagerie," is easily recognized by the closed, cage-like flowers with narrow slits perforating the sides. **Flower:** Lateral sepals are entirely fused and joined to the dorsal sepal at the base and tip. The result is an elongated flower resembling Aladdin's genie lantern in form, but with two slit-like windows that permit entry to pollinators. The column, lip, and small petals are concealed within. Ridged flowers are usually rich purple with darker splotches or stripes and a yellowish-orange interior; a few species are yellow or white. **Plant:** Lacking pseudobulbs, these small plants grow in clumps of clustered stems with funnel-shaped sheaths; some species have scrambling growth forms. Leaves are shorter than the stems, leathery and broad. Inflorescences of a single flower each emerge from the leaf bases.

Distribution and Diversity: Approximately 20 epiphytic species favor low- to middle-elevation forests. *Zootrophion* can be observed in the Caribbean and from Costa Rica to Bolivia and Brazil.

Ecology and History: Darwin studied *Z. atropurpureum* (when it was known as *Masdevallia fenestrata*, named for the window-like openings) during a period when he referred gleefully to his investigation of orchids as "a splendid sport" (Darwin 1862). He was perplexed by the flower's odd shape, musing, "the whole structure of the flower seems as if intended to prevent the flower from being easily fertilized" (Darwin 1877: 136). He eventually concluded that, "the presence of these two minute windows shows how necessary it is that insects should visit the flower" (137). Darwin subsequently discovered the same species at Kew Botanical Gardens with tiny eggs inside it, likely deposited by a small fly. His observation was confirmed recently in Costa Rica, where fly eggs were found in *Z. endresianum* flowers. Female flies enter through the lateral windows, searching for a suitable place for their eggs, a phenomenon known as brood-site deception. Presumably the confined space, protected from the elements and predators, makes an excellent nursery. Larger insects are prevented from entering by the size of the openings, a simple, mechanical form of pollinator selection and predator control.

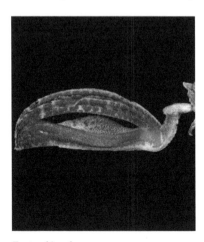  

Zootrophion dayanum
Ron Parsons

Z. serpentinum, showing broad leaves
Andy Phillips

Z. hypodiscus revealing yellow petals and lip
Andy Phillips

Zygopetalum

Description: *Zygopetalum* flowers are distinguished by greenish sepals and petals, both heavily marked with purple splotches, a spreading lip usually white with purple radiating stripes, and a pronounced yoke-like bulge on the lip's base.

Flower: Large (2–4" [5–10 cm]) flowers are usually strongly fragrant, with a scent resembling hyacinths. The genus name fuses the Greek words for "yoke" and "petal," referring to the raised, crescent-shaped bulge (callus), scored with parallel furrows like a plowed field. In some flowers, this callus better resembles the rear half of a horse saddle. The column is broad, robust, usually with a blunt yellow tip.

Plant: Most species have egg-shaped pseudobulbs (2–3" [5–7.5 cm] tall), becoming grooved with age, wrapped in brownish fibrous sheaths. Long (to 18" [46 cm]) leaves emerge from the pseudobulb tip and are deciduous in dry months. Leaves are narrow at the base, but broaden (to 2" [5 cm]) farther up and exhibit prominent veins. Inflorescences arise from the pseudobulb base bearing up to 8 flowers. **Similar:** See *Batemannia*, *Galeottia*, and *Promenaea*.

Distribution and Diversity: Some 15 species flourish terrestrially, or as epiphytes, in lower- and middle-altitude wet forests, from Venezuela to Peru and Brazil.

Ecology and History: *Zygopetalum* plants, sometimes called Ladybird Orchids, probably are visited by male euglossine bees seeking fragrances produced on the lip. The violet petal markings may help attract pollinators, as yet unidentified, that respond strongly to visual cues. *Zygopetalum maculatum* and *Z. crinitum* employ an uncommon strategy for improving reproductive efficiency: successfully pollinated flowers close slightly, indicating they are no longer receptive. These flowers, however, remain healthy and colorful for up to 3 months, whereas unpollinated flowers wilt and fall off after only 1 month. This tactic reduces the chance that a bee visits a pollinated flower, while simultaneously maintaining the magnetism of the overall floral display so that bees are attracted to unpollinated flowers. Unvisited flowers still may reproduce by means of apomixis, an asexual approach that yields seeds without fertilization, and offspring that are veritable clones of the parent. Evolution of these unique strategies may allow Zygopetalum to reproduce where suitable pollinators are rare, or absent entirely.

Zygopetalum maculatum
Joe Meisel

Rock-dwelling *Zygopetalum* in dry habitat
Ron Kaufmann

Z. maxillare showing leaves
Ron Kaufmann

Where to See Orchids

Nature Reserves and Conservation Sites

Orchids are arguably best viewed in their natural setting, where mossy branches and emerald fronds form perfect frames and backdrops. While you are enjoying wild orchids in bloom, you may happen upon a troop of monkeys, a flock of iridescent birds, or a stunning panorama—experiences that will remain with you forever. Numerous destinations throughout tropical America provide visitors the opportunity to appreciate orchids in their native habitat, from rainforest to cloud forest, and we urge you to undertake a trip to some of these sites. Local reserve managers and tourism operators earn their livelihood through such tours, and the income provided by visitors ensures that unspoiled orchid habitat will continue to be protected. We encourage you to use this guide to identify and learn more about the orchids you see, as well as to help plan your trip.

Below you will find a list of orchid destinations, organized (alphabetically) by country, for the region covered in this book (see map on page 218). They represent only a smattering of what each country has to offer, but primarily focus on orchid-rich habitats. Destinations ranging from national parks and private reserves to ecotourism lodges and botanical gardens are included. When you make travel plans to a particular country, we invite you to conduct further research to expand on this list. Contacting national or regional orchid societies can be an excellent starting place; members often know the best sites to find certain orchid species or habitat types and can give local advice on flowering seasons, climate, and other helpful tips.

The descriptions below are limited by space. We concentrate on the location of each site, including its elevational range, total land area, and the habitat types one may encounter there. Available amenities such as trails, lodging, and meals are listed, but not described; services not mentioned in the description are not currently available. For each site, basic directions for reaching its entrance are provided, although these are necessarily simplified. Contact information is restricted to official webpages, since telephone numbers and e-mail addresses may change over time. If the site lacks a website of its own, we suggest the reader use Internet search engines, which remain up-to-date, to obtain more information.

For city dwellers, orchids are more commonly encountered on tabletops or in florist shops, and although they are beautiful in those settings, the joy one experiences

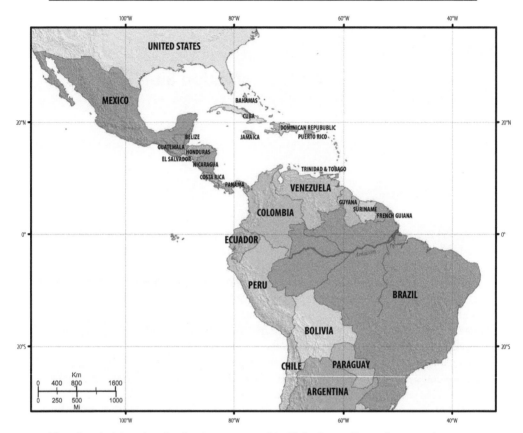

Map of tropical America, showing the area covered in this book as well as major mountain ranges and the Amazon River system

upon encountering a wild orchid in bloom is indescribable. Observing the flower in its natural environment, surrounded by the smells of the soil, the buzzing wings of potential pollinators, and the dripping of water from leaf tips, is richer in sensory detail than the flower shop can ever be. We invite you to indulge in this sensation by visiting one of the sites described here and discovering a flowering orchid yourself.

Conservation Sites by Country

Argentina

Iguazu National Park. More than 150,000 acres (67,000 ha) of lowland tropical forest (330 feet of elevation [100 m]) flank the world famous Iguazu Falls. Located in northeast Argentina, about 10 miles (16 km) from Puerto Iguazu, the park offers trails, meals, and a visitor center; lodging is available in a variety of nearby establishments. Website: www.iguazuargentina.com.

Yacutinga Lodge and Reserve. Lodging, meals, trails, and a visitor center are available in this small (1400 acres [570 ha]), privately operated reserve situated in lowland tropical forest. Yacutinga is located about 30 miles (50 km) upstream of Iguazu Falls (see above). Website: www.yacutinga.com.

The Bahamas

Rand Nature Center. A small but nationally supported sanctuary offers trails and a visitor center set in an attractive site. Its 100 acres (40 ha) showcase representatives of various island habitats. The center is found 2 miles (3 km) east of the center of Freeport. Website: use Internet search engines.

Belize

Belize Botanic Gardens. Visitors can enjoy several miles of trails, a visitor center, as well as food and lodging at this 45-acre (18 ha) private garden. Re-creations of savanna, rainforest, and other Belizean habitats are on display. Located 10 miles (16 km) from San Ignacio, the garden offers guided tours in English and Spanish. Website: www.belizebotanic.org.

Guanacaste National Park. Situated on the Belize River, this small (50 acre [20 ha]) national park offers hiking trails through lowland forest. Managed by the Belize Audubon Society, it lies some 50 miles (80 km) west of the capital of Belmopan. Website: search within www.belizeaudubon.org.

Bolivia

Amboro National Park. Lowland rainforest, mountain, and cloud forests are encompassed by this enormous park's 1.6 million acres (637,000 ha). Hiking trails and camping sites permit exploration of habitats ranging from 1100 to 11,500 feet of elevation (350–3500 m). Access is available from nearby towns of Buena Vista, Santa Fe, and Yapacani. Website: use Internet search engines.

Carrasco National Park. Trails, ranger stations, and campsites permit exploration of 1.5 million protected acres (622,000 ha). Upper elevation habitats including alpine grasslands cloak slopes ranging from 1000 to 15,000 feet (300–4700 m). The park is situated in the central state (department) of Cochabamba. Website: use Internet search engines.

Isiboro-Sécure National Park and Indigenous Territory. Mountainous terrain ranging from 600 to 9800 feet (180–3000 m) protects 2.9 million acres (1.2 million ha) of lowland and highland habitats. Hiking trails provide access from established ranger stations to a variety of forest types and campsites. The park is situated between the states of Beni and Cochabamba and can be visited from several entry points. Website: use Internet search engines.

Brazil

Augusto Ruschi Biological Reserve. Hillsides within the Atlantic Forest are protected within 8800 acres (3500 ha) that extend from 2600 to 3600 feet of elevation (800–1100 m). Trails and lodging are available, but special permission to visit this rainforest reserve must be obtained from the Brazilian Government (at present, permissions are managed by the Institute of Environmental and Renewable Natural Resources). Website: use Internet search engines.

Pseudolaelia vellozicola flowering on a granite formation just above the beach in eastern Brazil
Ron Kaufmann

Reserva Ecológica de Guapiaçu (REGUA). Encompassing more than 17,000 acres (7100 ha) of threatened habitat in the Atlantic Forest, the reserve is located just 50 miles (80 km) northeast of Rio de Janeiro. Visitors can obtain lodging and meals at the associated Guapi Assu Bird Lodge, which also maintains orchid lists. Habitats include lowland and montane forests, ranging from near sea level to 6500 feet of elevation (30–2000 m). Website: www.regua.co.uk.

Rio Atlantic Forest Trust (RAFT). Also located in the Atlantic Forest, this donation-supported conservation project protects two fragments of coastal forests rich in orchids. Several trails lead visitors from a lodge perched at 5000 feet of altitude (1500 m) through a variety of habitats and elevations. Website: www.riotrust.org.

Rio de Janeiro Botanical Garden. A nationally supported garden featuring orchids, other plants characteristic to the region, and more than 100 tropical bird species is set within 300 classically landscaped acres (140 ha). Located in the city of Rio itself, just west of the Lagoa (lagoon) Rodrigo de Freitas, the garden's paths are open daily to the public. Website: www.jbrj.gov.br, or use Internet search engines.

Colombia

Cali Botanical Garden. Located in the city of Cali, near the zoo, this garden affords visitors a unique experience in 30 acres (12 ha) of tropical dry forest along the Cali River. Hiking trails explore the grounds, guided tours are offered, dining services are available, and a variety of classes and events are regularly sponsored. Website: www.jardinbotanicocali.org.

El Refugio Torremolinos Reserve and Botanical Garden. Some 60 acres (25 ha) encompass native habitats and garden installations. The reserve, stretching between 5200 and 6200 feet (1600–1900 m), is dominated by cool, mountainous forests rich in orchids. Lodging and kitchen use are available at the biological station. Trails lead visitors through the forest and garden. Access is from Cali, about 14 miles (23 km) along the road to Dagua. Website: use Internet search engines.

Las Orquídeas National Park. Nearly 80,000 acres (32,000 ha) protect orchid-rich habitat ranging from tropical rainforest at 1100 feet (350 m) to cloud forest and alpine grassland at 11,000 feet (3400 m). Maintained trails and campsites provide access to the park, which is situated in the northwestern department (state) of Antioquia. Website: search national park site www.parquesnacionales.gov.co.

Tequendama Orchid Park (Parque Temático Orquídeas del Tequendama). Cloud forest is featured in this nongovernmental reserve located at 5400 feet of elevation (1650 m). Visitors can enjoy hiking trails and on-site meals; an orchid list is also available. The entrance is found in the town of Bellavista, just 45 minutes from Bogotá. Website: www.orquideas deltequendama.com.

Cattleya quadricolor blossoms bursting from a forest tree in the central Andes of Colombia
Ron Kaufmann

Yotoco Forest Reserve. Operated by the National University of Colombia, this Andean reserve protects 1380 acres (560 ha) of cloud forest between 4000 and 5200 feet (1200–1600 m). Food and lodging are available, as are trails, campsites, a science lab, and visitor center. The reserve is located 40 miles (65 km) north of Cali. Website: www.reservayotoco.palmira.unal. edu.co.

Costa Rica

National Park System. Costa Rica has established a superb collection of national parks, organized into a number of larger Conservation Areas throughout the country. The parks cover habitats ranging from lowland rainforest to seasonal dry forest to cloud forest and high alpine systems. Standouts include Braulio Carrillo (rainforest to cloud forest), Carara (rainforest and dry forest), Chirripó (cloud forest and alpine), Corcovado (rainforest and montane), La Amistad (rainforest and montane), Los Quetzales (montane forest), Palo Verde and Guanacaste parks (dry forest), Rincón de la Vieja (montane forest), and Volcán Poas (cloud forest). Orchid lists for many national parks are available (at www.epidendra.org). More information on these parks, and many others, can be obtained from the Costa Rican government (www.sinac.go.cr; in Spanish) and private sources (e.g., www.costarica-national parks.com).

Bosque de Paz (Peace Forest). Cloud forest rich in orchid diversity is the highlight of this approximately 2500-acre (1000 ha) site. Bird and orchid tourism are the focus of trails and guided tours through forests extending from 4800 to 8000 feet (1450–2450 m), connecting Poas Volcano and Juan Castro Blanco national parks. Meals and lodging, and even yoga classes, are available. The reserve is located northeast of San José, near the town of Palmira. Website: www.bosquedepaz.com.

Heliconias Lodge and Rainforest. Set on the doorstep of Tenorio Volcano National Park, this small, privately operated reserve offers trails, meals, and lodging. Lowland rainforest dominates the 140-acre (75 ha) property, located in northeast Costa Rica near the town of Bijagua. Website: www.heliconiaslodge.com.

Lankester Botanical Garden. Operated by the University of Costa Rica, the garden's 25 acres (10 ha) showcase a diverse mix of the country's incredible flora. The garden began life as an orchid collection (see genus account for *Chondrorhyncha*), and orchids remain the highlight of the collection. The entrance is in the town of Cartago, at 4500 feet (1370 m) of elevation, just 15 miles (25 km) east of the capital city San José. Website: use Internet search engines.

Las Alturas Biological Station. Rainforest and cloud forest are protected within this 25,000 acre (10,000 ha) mixed-use site extending from 3600 to 7200 feet of altitude (1100–2200 m). Trails provide access to the forest, which borders the international Parque Amistad (Peace Park); room and board are available. The privately operated station is located 20 miles (30 km) northeast of San Vito. Website: www.lasalturas.com.

Las Cruces and La Selva Biological Stations. Both sites are operated by the Organization for Tropical Studies, a consortium of universities that promote research and conservation. Stations are open to the public and offer trails, food, and lodging, as well as extensive lists of fauna and flora, including orchids. Las Cruces is located near the Panama border and encompasses more than 800 acres (325 ha) of montane forest between 3300 and 4600 feet (1000–1400 m). La Selva is situated in the eastern lowlands, between 120 and 9500 feet (36–2900 m), and protects nearly 4000 acres (1600 ha) of rainforest and montane forest. Website: search within www.ots.ac.cr.

Monteverde Cloud Forest Reserve. World-famous destination featuring 35,000 acres (14,200 ha) of pristine montane and cloud forest, at an elevation of 4600 feet (1400 m). Trails, lodging, meals, guides, orchid lists, and numerous other services are available in the reserve and surrounding communities. Monteverde is located just a few minutes east of the town of Santa Elena and is the centerpiece of a veritable smorgasbord of ecotourism destinations in this attractive mountainous region. Website: www.monteverdeinfo.com.

Santa Elena Cloud Forest Reserve. A neighbor of Monteverde, protecting 760 acres (310 ha) of middle-elevation cloud forest (5000 feet [1500 m]), this private reserve offers trails, a visitor center, and meals. Lodging is available in the nearby town of Santa Elena. Website: www.reservasantaelena.org.

Cuba

Jardín Botánico Orquideario de Soroa (Soroa Orchid Botanical Garden). Located inside the Sierra del Rosario Biosphere Reserve (see below), the garden's 85 acres (35 ha) and greenhouses feature an extensive orchid collection. Access to the garden, situated at 675 feet (206 m) above sea level, is from the town of Soroa, 50 miles (85 km) west of Havana. Website: use Internet search engines.

National Botanic Garden (Jardín Botánico Nacional). Located at sea level, on nearly 1500 acres (600 ha), the garden presents visitors with representatives of a variety of Cuban habitats. A functioning research site, the garden focuses on environmental education and the conservation of plants endemic to the island. The entrance is 10 miles (15 km) from the center of Havana. Website: www.uh.cu/centros/jbn/.

Sierra del Rosario Biosphere Reserve. Tropical dry forest is featured in this 66,000-acre (26,686 ha) reserve that ranges from 150 to 1800 feet (50–550 m). Visitors will find hiking trails, an information center, and orchid lists; lodging is available. Access is from the Las Terrazas entrance 45 miles (70 km) west of Havana, or from the Soroa entrance 50 miles (85 km) west of Havana. Website: use Internet search engines.

Curaçao

Christoffel National Park. Hiking trails offer access to some 4000 acres (1800 ha) of varied island habitats, spanning elevations from sea level to 1200 feet (380 m). A visitor center is open to the public at the entrance, just a few minutes north of the town of Barber. Website: www.christoffelpark.org.

Dominican Republic

Armando Bermudez National Park. A range of habitat types, from rainforest to pine forest, can be visited within the park's nearly 200,000 acres (79,000 ha). Lodging is available, and trails provide access to elevations ranging from 2950 to 10,100 feet (900–3080 m). The main park office is located in La Ciénaga. Website: use Internet search engines.

El Choco National Park. Lowland tropical rainforest, scattered lagoons, and some 200 caves are highlights in this 19,000 acre (7700 ha) protected area. Walking trails, canoes, and horses permit access to the park, and guided tours are available; lodging and meals can be found in the nearby town of Cabarete. Website: use Internet search engines.

José del Carmen Ramírez National Park. A ranger station and rustic bunkhouses provide a jumping-off point for trails that explore 1900 acres (764 ha) of forest. The park lies astride the border between San Juan and Santiago provinces. Website: use Internet search engines.

National Botanic Garden (Jardín Botánico Nacional Dr. Rafael Ma. Moscoso). The national garden features representations of various island habitats within a 500-acre (200 ha) site, as well as permanent displays of orchids, bromeliads, and ferns. Trails, meals, a visitor center, and an orchid list are available. The garden is located in the capital city of Santo Domingo (245 feet of elevation; 75 m). Website: www.jbn.gob.do.

Sierra del Bahoruco National Park. Middle-elevation park offers visitors the chance to explore 200,000 acres (80,000 ha) of humid, seasonally dry, and montane tropical forest rich in endemic plants. This large park climbs from 4800 to 6900 feet (1460–2100 m). A list of orchid species is available. Visitor entrance is located approximately 12 miles (20 km) northwest of Barahona. Website: use Internet search engines.

Ecuador

National Park System. Ecuador has an extensive and well-developed system of national parks and ecological reserves. Among the most well-known are the Galápagos Islands and rainforest reserves such as Yasuní National Park. Other sites provide visitors with access to millions of acres of mountainous habitat, including cloud forest and tropical alpine grasslands. These upper-elevation destinations include Cajas, Cayambe-Coca, Cotopaxi, Podocarpus, Sangay, and Sumaco parks. All are open to the public, with access provided by road, hiking trail, and horseback, depending on the site. More information can be obtained from the Ecuadorian Ministry of the Environment (www.ambiente.gob.ec; Spanish only) and from individual park webpages (use Internet search engines).

Bellavista Cloud Forest Reserve. Visitors can enjoy comfortable accommodations and trails that explore 1700 acres (700 ha) of Andean montane forest. Orchids are plentiful in this elevational range of 5000 to 8200 feet (1550–2500 m). The reserve and lodge are located about 2 hours northwest of Quito, mostly along the road to Mindo. Website: www.bellavistacloudforest.com.

Bosque Seco Lalo Loor (Lalo Loor Dry Forest). Some of the last examples of coastal semi-deciduous forest are protected in this 500-acre (200 ha) reserve. The site offers an educational visitor center, extensive trail system, accommodations, and meals. Lists of orchids, birds, and mammals are available, as are guide services. The reserve, managed by the Ceiba Foundation for Tropical Conservation, is located just south of the equator, between the coastal cities of Pedernales and Jama, in northern Manabí province. Website: www.ceiba.org.

Cerro Candelaria Reserve. Over 6000 acres (2600 ha) of montane forest and alpine grassland protect excellent orchid habitat. Access to the reserve, which ranges from 5900 to 12,600 feet in altitude (1800–3860 m), must be arranged by contacting the EcoMinga Foundation. Website: www.ecominga.com.

Ecuagenera. This established commercial orchid nursery also offers orchid tourism packages that travel throughout Ecuador, leading visitors to dozens of sites featuring abundant wild orchids. The organization maintains two private reserves totaling 750 acres (300 ha) at elevations from 6900 to 10,500 feet (2100–3200 m). Tours include all lodging, travel, meal, and guiding costs. The main nursery is located in Gualaceo, 20 minutes east of Cuenca. Website: www.ecuagenera.com.

El Monte Ecolodge. Visitors to this middle-elevation (4600–8500 feet [1400–2600 m]) site can enjoy trails through 100 acres (44 ha) of cloud forest. Meals and accommodations are available at the lodge, which also operates the Mindo Biological Station. Both are located 2 miles (3 km) outside the town of Mindo. Website: www.ecuadorcloudforest.com.

El Pahuma Orchid Reserve. A conservation easement, the first of

Misty cloud forests of the El Pahuma Orchid Reserve give a mossy home to *Oncidium cirrhosum* and hundreds of other orchids
Joe Meisel

its kind in South America, protects 1500 acres (650 ha) of pristine cloud forest. More than 300 species of orchids are estimated to occur in the reserve, which ranges in elevation from 6200 to 8500 feet (1900–2600 m). Locally owned and operated, the reserve is a project of the Ceiba Foundation for Tropical Conservation. Hiking trails, a visitor center, restaurant, and lodging are available, as well as a maintained orchid garden highlighting local species. The reserve is located 45 minutes northwest of Quito, between the towns of Calacalí and Nanegalito, on the road to Mindo. Website: www.ceiba.org.

Jardín Botánico Las Orquídeas (The Orchids Botanical Garden). Located in the eastern foothills of the Andes at 3100 feet of elevation (950 m), this sprawling garden offers some 17 acres (7 ha) of botanical displays including an ample collection of orchids. They also engage in scientific research, habitat restoration, and local conservation programs. The entrance is 2 miles (3 km) southeast of Puyo. Website: www.jardinbotanicolasorquideas.com.

Los Cedros Biological Reserve. Operated by a local nonprofit organization (Fundación Los Cedros), the reserve protects 17,000 acres (6880 ha) of Andean cloud forest. Trails provide access to habitat ranging from 3300 to 8800 feet in altitude (1000–2700 m). Meals and lodging are available, as is an orchid list. The entrance is located 40 miles (60 km) northwest of Quito, but visits must be pre-arranged. Website: www.reservaloscedros.org.

Mindo Orchid Garden. Situated in the heart of the popular ecotourism town of Mindo (5200 feet of elevation [1600 m]), this botanical garden features more than 200 orchid species in its collection. Lodging and meals are provided at any of the numerous lodges in town, located approximately one and a half hours northwest of Quito. Website: use Internet search engines.

San Antonio Cloudforest Reserve. Located in the Pastaza River Valley, this privately owned reserve protects 200 acres (80 ha) of wet, mountainous forest between 7900 and 10,800 feet (2400–3300 m). Lodging, meals, and guide services to the reserve can be arranged in the nearby town of Baños. Website: www.lunaruntun.com.

El Salvador

El Imposible National Park. Administered by a local nongovernmental organization (SalvaNatura), this park protects 9400 acres (3820 ha) of foothill and dry forest. Trails provide access to hilly forests ranging in altitude from 800 to 4700 feet (250–1425 m). Meals, accommodations, and a visitor center are available to the public. The park is located in the southwest corner of the country, with entry from the town of Cara Sucia. Website: www.salvanatura.org.

La Laguna Botanical Garden. Most of the site's 70 acres (30 ha) are dedicated to protection of primary forest habitat; the remainder have been developed as a classical botanical garden with collections that include orchids, bromeliads, and numerous other tropical plants. A visitor center and meals are available. The site, located at 2600 feet above sea level (800 m), is situated south of Antiguo Cuscatlán. Website: www.bgbm.org/lagu/JardinBotanico.htm.

Los Volcanes (The Volcanoes) National Park. Visitors can hike through habitats including montane forest, alpine grasslands, and old lava flows within the park's 11,000 acres (4500 ha). The protected area stretches from 1600 to 7800 feet (500–2380 m) of elevation and is located a few minutes north of Sonsonate. Website: www.salvanatura.org.

Montecristo National Park. Hiking and camping allow visitors to enjoy some 5000 acres (2000 ha) of forest that straddles the borders between El Salvador, Honduras, and Guatemala. Dry forest, mixed pine-oak stands, and cloud forest are encountered as the park ascends from 2600 to 7900 feet (800–2400 m). The entrance is near Metapan, about 4 hours north of San Salvador. Website: use Internet search engines.

Guatemala

Biotopo Cerro Cahui. Located within the enormous Maya Biosphere Reserve, this protected area comprises 1700 acres (700 ha) of tropical rainforest. Visitors can enjoy hiking trails and camping within the lowland site (330–1200 feet of elevation [100–360 m]). The reserve entrance is near El Remate, in the extreme northeast of the country. Website: use Internet search engines.

Chelemhá Cloud Forest Reserve. Trails and comfortable accommodations permit visitors to enjoy 1250 acres (500 ha) of lush cloud forest. The reserve ranges from 6200 to 8300 feet in elevation (1900–2530 m) and is located 18 miles (26 km) north of Tucurú. Website: www.chelemha.org.

El Mirador National Park. A large park (290,000 acres [117,000 ha]) situated on the borders with Mexico and Belize, this reserve protects lowland rainforest and wetland habitats. Hiking and camping are permitted. Entry is from the ranger station 40 miles (60 km) north of Carmelita. Website: use Internet search engines.

Quetzal Biotope (Mario Dary Rivera Nature Reserve). Middle-elevation montane forest (5200–7500 feet [1600–2300 m]) presents hikers and campers with nearly 3000 acres (1200 ha) to explore. Lists of orchids found in the reserve are available at the visitor center. Access is from the village of Purulhá, 100 miles (160 km) from Guatemala City. Website: use Internet search engines.

Tikal National Park. In addition to its world-renowned Mayan ruins, this sprawling park conserves more than 142,000 acres (57,600 ha) of tropical rainforest rich in flora and fauna. Three hotels and a camping area are available within the park, as is a visitor center. The park can be visited by car or bus from Flores and El Remate where meals and additional lodging can be found. Website: www.tikalinformation.com.

Honduras

Celaque National Park. More than 65,000 acres (26,600 ha) of protected habitat include pines, deciduous forest, and cloud forest. Numerous villages exist within the park, where lodging and meals can be obtained. Hiking trails permit access to some of the park's elevation range, from 3200 to 9350 feet (975–2850 m). The entrance is 5 miles (8 km) from the town of Gracias. Website: use Internet search engines.

Cusuco National Park. Drier habitats, including deciduous and pine forests, await visitors hiking in the 57,800 acre (23,400 ha) reserve. Upper elevations (to 79,500 feet [2425 m]) support cloud forest formations. Overnight lodging is available, as is a visitor center. The park is situated just 25 miles (40 km) west of San Pedro Sula. Website: use Internet search engines.

Montaña de Yoro National Park. This large protected area encompasses 38,000 acres (15,366 ha) of montane and elfin forest. Trails and overnight lodging allow exploration of a park that extends in elevation from around 5900 to 7360 feet (1800–2245 m). Entry is located near the town of Yoro. Website: use Internet search engines.

Pico Bonito National Park. Nearly 140,000 acres (56,430 ha) of habitat, from lowland rainforest to cloud forest, are protected. Trails are provided, although lodging and meals are not available. The park ranges from 200 to 8100 feet (60–2480 m) of altitude. Entrances are located at El Pino and Rio Cangrejal, near the city of La Ceiba. Website: use Internet search engines.

Pico Pijol National Park. Access to the rugged mountains of the 30,000-acre (12,200 ha) park is only possible with guides, as formal trails are lacking. The upper elevation core zone (5900 to 7500 feet [1800–2280 m]) harbors cloud forest on steep, epiphyte-rich slopes. Explorations are best begun in the town of Morazán. Website: use Internet search engines.

Rio Plátano Biosphere Reserve. A coastal site protecting nearly 1.3 million acres (525,000 ha) of rainforest, this reserve can be visited only by organized tour, as no roads reach its remote borders. Occupying the Mosquito Coast plain, the park's interior is mountainous, reaching 4900 feet (1500 m). Tours depart from La Ceiba. Website: search within whc.unesco.org.

Sierra de Agalta National Park. Very rugged terrain is characteristic of steep forests that rise from 5900 to 7700 feet (1800–2350 m). Hiking and camping are permitted in the reserve's 51,300 acres (20,785 ha) of montane forest. The park entrance is 6 miles (10 km) southwest of Gualaco. Website: use Internet search engines.

Jamaica

Cinchona Botanical Gardens. Set in a private farm at 3500 feet (1000 m) of elevation in the Blue Mountains, the gardens present visitors with an array of plants native to the island, including orchids. Lodging and meals are available, as are hiking trails into the mountains. The entrance is 1.5 hours east of Kingston by four-wheel-drive vehicle; transport can be arranged by the farm. Website: search within www.limetreefarm.com.

Mexico

El Cielo Biosphere Reserve. This expansive protected area in the Sierra Madre mountains comprises 350,000 acres (144,000 ha) of rainforest, pine-oak formations, cloud forest, and alpine terrain. Stretching from 650 to 7500 feet above sea level (200–2280 m), El Cielo encompasses a mosaic of land uses and many local inholdings. Trails and a community-operated lodge are available to visitors.

Located south of Ciudad Victoria, the reserve can be entered from the villages of Alta Cima and San José. Website: use Internet search engines.

Jardín Botánico Orquídeas Moxviquil (Orchid Botanical Garden). Situated within a 250-acre (100 ha) forest reserve is a small (5 acre [2 ha]) but energetic botanical garden project dedicated to the rescue and preservation of Chiapas' orchids. More than 400 species of orchids are in the collection, and other programs focus on promotion of local orchid societies and environmental education programs. The garden is located in San Cristobal de las Casas. Website: www.orchidsmexico.com.

Los Tuxtlas Biosphere Reserve. Located in the eastern state of Veracruz, this coastal reserve protects more than 380,000 acres (155,000 ha) and includes volcanic peaks reaching 5500 feet of altitude (1700 m). Habitat types include rainforest, cloud forest, and mountain lakes. Access is from the village of Sihuapan, where meals and other services can be found. Website: use Internet search engines.

Sierra Gorda Biosphere Reserve. Operated through a community-based partnership, the reserve encompasses nearly one million acres (383,000 ha). Core areas protect semi-dry forest types including pines, oaks, and magnolias; numerous caves also dot the region's karst topography. The reserve rises from a low point of 1000 feet (300 m) to peaks reaching 10,100 feet (3100 m), and can be reached from Jalpan de Serra, 80 miles (130 km) northeast of Santiago de Querétaro. Website: use Internet search engines.

Vallarta Botanical Gardens. This sprawling, 20-acre (8 ha) garden was developed in a mix of open and semi-forested settings, and boasts a splendid collection of Mexican orchid species. Pines and deciduous forest dominate the site, located at 1300 feet (400 m) of altitude. Ongoing projects include reforestation, research, and environmental education. Meals are available in the visitor center. The garden entrance is some 30 minutes south of the city of Puerto Vallarta. Website: www.vbgardens.org.

Nicaragua

El Jaguar Cloud Forest Reserve. The Isabelia Mountain Range is home to this small (200 acre [80 ha]), privately operated reserve. At 4400 feet (1350 m) above sea level, it protects a mixture of organic coffee fields and montane forest, both fine systems for orchid diversity and migratory birds. Lodging and meals are available at a modest biological station. The reserve entrance is 17 miles (27 km) outside of Jinotega, north of Managua. Website: www.jaguarreserve.org.

Mombacho Cloud Forest Reserve. Several trails lead visitors up the flanks of Mombacho Volcano, where the reserve protects 2500 acres (1000 ha) of hillside and cloud forest between 130 and 4400 feet of altitude (40–1345 m). Camping is allowed, and a small biological station provides indoor lodging and meals. The entrance is just 6 miles (10 km) south of Granada. Website: www.mombacho.org.

Selva Negra Cloud Forest Reserve. More than 1800 acres (750 ha) of cloud forest and bird-friendly coffee plantations are situated at 3900 feet (1200 m) above sea level. Lodging and meals are available, and trails lead visitors through good orchid habitat. Access is 15 minutes north of Matagalpa. Website: www.selvanegra.com.

Panama

APROVACA Orchid Conservation Center. Formed by orchid producers in nearby communities, the Center focuses on protection of Panama's threatened orchid species. Propagated orchids are re-introduced into forests of the nearby Cerro Gaital Natural Monument. Overnight accommodations, meals, trails, and a visitor center are available. The site is located in the village of El Valle, about 75 miles (120 km) west of Panama City. Website: www.aprovaca.org.

Peristeria elata, nicknamed the Dove Orchid for its resemblance to a pigeon in a nest, is the national flower of Panama
Ron Kaufmann

Barro Colorado Island Research Station. Formerly a mountaintop, the island was created when the surrounding lowlands were flooded to fill the Panama Canal; today its 3700 acres (1500 ha) of tropical rainforest are managed as a world-class research site by the Smithsonian Tropical Research Institute. Visits to the island, which has an excellent trail system through wildlife and plant-rich habitat, can be arranged through the Smithsonian. Access is from Panama City. Website: search within www.stri.si.edu.

Baru Volcano National Park. A lofty, inactive volcano with seven craters is swaddled by this large park (35,000 acres [14,320 ha]). The foothills are dominated by lower montane forest, the 11,400-foot summit (3475 m) by cloud forest. Hiking trails permit visitors to view quetzals and orchids, but campsites are the only lodging within the park. The entrance and ranger station is just outside Boquete. Website: use Internet search engines.

Finca Drácula. This privately operated reserve and lodge is set on 30 acres (12 ha) at 6500 feet (2000 m), adjacent to the enormous La Amistad Biosphere Reserve. Visitors can enjoy lodging, meals, trails, and a large orchid collection (more than 2200 species including, not surprisingly, many species of *Dracula*). The lodge is located in Guadalupe, 45 miles (70 km) north of the city of David. Website: www. fincadracula.com.

La Fortuna Forest Reserve and Research Station. Operated by the Smithsonian Tropical Research Institute, the station is located in a hydrological reserve protecting 48,000 acres (19,500 ha) of montane forest. Visitors can arrange lodging in cabins, use a communal kitchen, and explore trails passing through middle-elevation (6560 feet [2000 m]) mountain forest. The station is just 6 miles (10 km) west of David. Website: search within www.stri.si.edu.

Los Quetzales EcoLodge. A high-elevation retreat offering meals, lodging, and trails through some 990 acres (400 ha) of cloud forest. Situated at 7200 feet (2200 m), the property also abuts the million-acre (407,000 ha) La Amistad reserve. The lodge is reached from Guadalupe, about an hour drive north of the town of David. Website: www.losquetzales.com.

Paraguay

Cerro Corá National Park. Located near the border with Brazil, this mixed-habitat park protects some 30,000 acres (12,000 ha) of dry tropical forest and savanna, as well as several caves decorated with pre-Colombian art. Immense rock outcroppings reach higher than 1000 feet (320 m) of elevation. A visitor center is open to the public, and camping, cabins, and guide services are available. The park entrance is located near the border town of Pedro Juan Caballero. Website: use Internet search engines.

Ñacunday National Park. The impressive Ñacunday Falls (130 feet high, 360 feet wide [40 and 110 m, respectively]) form the centerpiece of this 5000-acre (2000 ha) forest reserve. Flowing into the Parana River that forms the border with Argentina, the falls are surrounded by lowland forest (400 feet of elevation [120 m]) and riparian zones harboring considerable orchid diversity. Trails permit visitor access to the falls and forest. The park entrance is some 50 miles (80 km) south of Ciudad del Este. Website: use Internet search engines.

Peru

Machu Picchu National Sanctuary. Surrounding the world-famous Incan ruins, perched at 7900 feet above sea level (2430 m), are some 94,000 acres (38,000 ha) of protected cloud forests, river gorges, and alpine grasslands. Visitors to the reserve may arrive by bus, train, horseback, or by hiking the renowned Inca Trail. Taking time to explore trails through nearby natural areas will provide fantastic orchid-viewing opportunities. Numerous hotels offer meals and accommodations, as well as guide services. Website: use Internet search engines.

Manu National Park and Biosphere Reserve. Combined, these two protected areas encompass several million acres of pristine Amazon rainforest, as well as local escarpments and ridges supporting montane and cloud forest formations. Some of the finest and most biodiverse tropical forests on Earth can be found in these reserves. Access is strictly controlled, and most visitors choose to stay at one of the many lodges located on or near the borders. Just four such sites are highlighted below, but many other similar destinations are available.

Cock of the Rock Lodge. Named for the brilliant red-orange birds that can be observed here, this lodge offers meals, accommodations, and trails through 12,500 acres (5060 ha) of montane and cloud forest (5250 feet [1600 m]). Situated within the Manu Biosphere Reserve, the lodge entrance is on the highway connecting Cuzco and Shintuya. Website: use Internet search engines.

Manu Learning Centre. Established adjacent to Manu National Park, the station offers accommodations, meals, science labs, and trails providing access to its own 1580 acre (640 ha) protected area. Guided tours take visitors on multi-day excursions into the park, which comprises some 4 million acres (1.6 million ha). Access is by boat from the port town of Atalaya. Website: use Internet search engines.

Manu Wildlife Center. Comfortable lodging greets visitors to the center, built in a privately managed 40,000-acre (16,190 ha) forest bordering Manu National Park. Guided tours are provided through unparalleled lowland Amazonian rainforest (1000 feet [300 m]). Access to the site is by boat from Puerto Maldonado. Website: use Internet search engines.

Posada San Pedro Lodge. Highlighting the upper elevation habitats of the Manu region, this destination is perched in cloud forest at an altitude of 5250 feet (1600 m). Meals and lodging are provided, and visitors may choose from several trails winding through the surrounding forest. The lodge is a 6-hour drive from Cusco. Website: use Internet search engines.

Owlet Lodge. The Abra Patricia-Alto Nieva conservation area in northern Peru, in which the lodge is located, protects more than 24,000 acres (9700 ha) of cloud forest. The region is famed as one of the country's finest birding destinations; pristine upper-elevation forest (7600 feet [2300 m]) also harbors an impressive diversity of orchids. Accommodations, meals, guides, and hiking trails are available. The lodge is located some 5 hours from Tarapoto. Website: use Internet search engines.

Puerto Rico

El Yunque National Forest Reserve. Significant research on rainforest and montane forest ecology, including many studies on orchids, has been conducted in this 28,000-acre (11,330 ha) reserve since its establishment. Lodging is available in and near the reserve, and camping within the forest is allowed (permit required). Orchid lists can be obtained, and numerous hiking trails permit exploration of the Luquillo Mountains, which reach 3500 feet of elevation (1060 m). Located near the eastern tip of the island, the reserve entrance is close to the town of La Vega, on highway 191, some 40 miles (65 km) east of San Juan. Website: www.fs.usda.gov/elyunque.

San Juan Botanical Garden. Situated in the south side of San Juan, this formal garden is operated by the University of Puerto Rico. Its 300-acre (120 ha) grounds showcase various habitats containing more than 30,000 plants, including numerous orchid species. The garden entrance, situated in Río Piedras, is open daily. Website: search within www.upr.edu.

Toro Negro State Forest Reserve. Rainforest and montane forest both can be explored within the 6900 acres (2800 ha) of protected area. Ranging from 1400 to 4400 feet (440–1340 m), the reserve offers trails winding through fine forests and past impressive waterfalls. The entrance is located a few minutes north of Villalba. Website: use Internet search engines.

Suriname

Central Suriname Nature Reserve. A World Heritage Site, the reserve protects 4 million acres (1.6 million ha) of lowland rainforest and montane forest, and numerous granite outcroppings (inselbergs). Stretching from 80 to 4000 feet in elevation (25–1230 m), much of the massive reserve remains formally unexplored.

Visitors must arrive by canoe or airplane, as no roads serve the area. Lodging is available at several park airstrips, and a number of local tour agencies offer multi-day guided excursions. Website: use Internet search engines.

Venezuela

Canaima National Park. Home to many of the stunning and biologically diverse tepuis (tabletop mountains), this park comprises 7.5 million acres (3 million ha) of savannas and rainforest punctuated by the soaring mesas. Most famous among the tepuis are Mount Roraima (the easiest to scale) and Ayantepui (site of Angel Falls, the world's tallest waterfall at 3200 feet [979 m]). The park typically is reached by airplane, but roads from Cuidad Bolivar provide access to the western sector. Website: use Internet search engines.

Delta del Orinoco National Park. Established thanks to vigorous lobbying by renowned orchid expert Julian Steyermark (see genus account for *Bifrenaria*), the park protects 820,000 acres (331,000 ha) of lowland forest and winding waterways of the Orinoco River. Access can be arranged through tour operators. The park is situated on the far northeast coast. Website: use Internet search engines.

Henri Pittier National Park. The country's first national park, named for the talented Swiss botanist who worked extensively in the region, encompasses 266,000 mountainous acres (107,800 ha) ranging from sea level to 8000 feet (2430 m). Lodging is not provided within the park but is available in nearby communities. Hiking trails and a ranger station facilitate exploration of rainforest, dry forest, mangroves, and cloud forest. Roads linking the city of Maracay to the coast pass through the park and provide access. Website: use Internet search engines.

Rancho Grande Biological Station. Established as a research site, the station is located within Henri Pittier National Park and is open to the public. Trails depart from cabins situated at 3600 feet (1100 m) above sea level and enter the immense park. The station entrance is 20 miles (30 km) from Maracay on the road to Ocumare. Website: use Internet search engines.

U.S. Virgin Islands

St. George Village Botanical Garden. On the island of St. Croix, this sea-level garden maintains representations of island ecosystems and a collection of native plant species in a 16-acre (6.5 ha) site. Paths lead visitors through educational displays and collections including orchids and bromeliads. The garden entrance is 2 miles (3 km) east of Frederiksted. Website: www.sgvbg.org.

Virgin Islands National Park. More than 14,700 acres (5960 ha) of coastal and montane forest, and fine snorkeling reefs, are protected on the island of St. John. The park ranges from sea level to over 1270 feet in elevation (389 m). Hiking trails, including a self-guided trail, ranger-guided programs, and the park visitor center are available to tourists. Camping is an option, or lodging can be obtained on the island beyond park boundaries. St. John Island can be reached by boat only; ferry service is available. Website: search within www.nps.gov.

Bibliography

Online Resources

Taxonomy Websites

Kew Royal Botanical Gardens world checklist of selected plant families
http://apps.kew.org/wcsp/home.do
Enormous searchable database of the current taxonomic status of plants in more than 170 families, including an exhaustive treatment of orchids

The Plant List
http://www.theplantlist.org
Exhaustive database of more than one million plant names, including orchids, with detailed taxonomic information and links to primary taxonomic records

eMonocot
http://e-monocot.org
Collaborative effort among monocot taxonomists, providing tools for identification and classification of plants including orchids

Epidendra global orchid taxonomic network
http://epidendra.org
Extensive searchable orchid database of orchid taxonomy, managed by University of Costa Rica and Lankester Botanical Garden, including links to literature and scanned herbarium specimens

Harvard University Herbarium databases
http://kiki.huh.harvard.edu/databases
Searchable indices of plant names, botanists, botanical publications, and herbarium specimens

International Plant Names Index
http://www.ipni.org
Searchable database of scientific plant names, including orchids

University of Missouri Botanical Garden Tropicos database
http://tropicos.org
Access to Missouri's unparalleled collection of tropical botanical records, providing taxonomic and conservation information on plants, including orchids

Photograph Websites

Internet Orchid Species Photo Encyclopedia
http://www.orchidspecies.com
One of the most comprehensive sources for photographs of more than 15,000 species of orchids, also providing text descriptions for each species and information on geographic range, flower size, and links to taxonomic sources

Eric Hunt orchid species photographs
http://www.orchidphotos.org/images/orchids/index.html
Large collection of high-quality photographs of orchid species, arranged by genus

London, Ontario, Orchid Society picture search
http://los.lon.imag.net/picref.asp
Extensive searchable database of books and other printed material in which to find orchid photographs and line drawings

Miscellaneous Websites

American Orchid Society
http://www.aos.org
Home page of the preeminent orchid society in the United States, providing broad information on orchid taxonomy, cultivation, conservation, and research

Swiss Orchid Foundation BibliOrchidea
http://orchid.unibas.ch/bibliorchidea.index.php
Enormous searchable database of orchid literature; access requires free membership

Brazilian Orchids
http://delfinadearaujo.com
Wide range of resources specifically for Brazilian orchids, including a substantial photo gallery, geographically organized lists of orchids, and links to local orchid societies

Orchid Board community forum
http://www.orchidboard.com/community
Message forum for orchid enthusiasts to post questions and answers about a wide range of tropics, from cultivation advice to species identification

The Orchid Mall
http://orchidmall.com
Collection of links to a variety of orchid resources including cultivation supplies, public speakers, plant vendors, and a wide range of publications

Orchidwire
http://www.orchidwire.com
Collection of links to hundreds of orchid associations, plant vendors, and other online resources

Orchid Digest quarterly journal
http://orchiddigest.com
One of America's most photographically arresting orchid journals, providing excerpts from current and back issues covering a wide range of orchid topics, from taxonomy to natural history

Publication Resources

Ackerman, J.D. 1983. Euglossine bee pollination of the orchid, *Cochleanthes lipscombiae*: a food source mimic. *Amer. J. Bot.* 70(6): 830–834.

Ackerman, J.D., J.A. Rodriguez-Robles, and E.J. Meléndez. 1994. A meager nectar offering by an epiphytic orchid is better than nothing. *Biotropica* 26(1): 44–49.

Aliscioni, S.S., J.P. Torretta, M.E. Bello, and B.G. Galati. 2009. Elaiophores in *Gomesa bifolia* (Sims) M.W. Chase & N.H. Williams (Oncidiinae: Cymbidieae: Orchidaceae) structure and oil secretion. *Ann. Bot.* 104(6): 1141–1149.

Almeida, A.M., and R.A. Figueiredo. 2003. Ants visit nectarines of *Epidendrum denticulatum* (Orchidaceae) in a Brazilian rainforest: effects on herbivory and pollination. *Braz. J. Biol.* 63(4): 551–558.

Alrich, P., and W. Higgins. 2008. *The Marie Selby Botanical Gardens Illustrated Dictionary of Orchid Genera.* Ithaca, NY: Cornell University Press, in association with Selby Botanical Gardens Press, Sarasota, FL.

Ames, O. 1948. *Orchids in Retrospect: A Collection of Essays on the Orchidaceae.* Cambridge, MA: Botanical Museum of Harvard University.

Ames, O., and D.S. Correll. 1962. *Orchids of Guatemala.* Chicago, IL: Chicago Natural History Museum.

Anonymous. 1898. Obituary of Mr. John Weir. *Gardener's Chronicle* 23:301.

Arditti, J. 1967. Factors affecting the germination of orchid seeds. *Botanical Review* 33(1): 1–97.

Arditti, J. (ed.). 1984. *Orchid Biology, Reviews and Perspectives,* vol. 3. Ithaca, NY: Cornell University Press.

Arditti, J. 1992. *Fundamentals of Orchid Biology.* Ann Arbor, MI: University of Michigan Press.

Arditti, J. (ed.). 1994. *Orchid Biology, Reviews and Perspectives,* vol. 6. Ithaca, NY: Cornell University Press, Ithaca.

Balfour, A. 1867. Obituary notices of James Smith, Esp. of Jordanhill; of Dr. G.A. Martin, Isle of Wight; and of George Ure Skinner, Esq. of Guatemala. *Transcripts of the Botanical Society of Edinburgh* 9:85.

Barreto, D.W., and J.P. Parente. 2006. Chemical properties and biological activity of a polysaccharide from *Cyrtopodium cardiochilum. Carbohydrate Polymers* 64(2): 287–291.

Barringer, K. 1985. Three new species of *Elleanthus* (Orchidaceae) from Central America. *Brittonia* 37(3): 286–290.

Beattie, A.J. 1985. *The Evolutionary Ecology of Ant-Plant Mutualisms.* Cambridge, UK: Cambridge University Press.

Benitez-Vieyra, S., A.M. Medina, E. Glinos, and A.A. Cocucci. 2006. Pollinator-mediated selection on floral traits and size of floral display in *Cyclopogon elatus*, a sweat bee-pollinated orchid. *Func. Ecol.* 20(6): 948–957.

Bentley, B.L. 1977. Extrafloral nectaries and protection by pugnacious bodyguards. *Annual Rev. Ecol. Syst.* 8:407–427.

Berliocchi, L. 1996. *The Orchid in Lore and Legend.* Portland, OR: Timber Press.

Blanco, M.A. 2002. Notes on the natural history of *Cyclopogon obliquus* (Orchidaceae: Spiranthinae) in Costa Rica. *Lankesteriana* 5:3–8.

Borrell, B.J. 2007. Scaling of nectar foraging in orchid bees. *American Naturalist* 169(5): 569–580.

Botanical Society of Edinburgh. 1891 to 1970. *Transactions and Proceedings of the Botanical Society of Edinburgh,* vol. 9. Edinburgh, Scotland.

Botanischer Garten München-Nymphenburg. 2012. The genus *Coryanthes.* http://www.botmuc.de/en/about/guenter_gerlach/genus_coryanthes.html (accessed 2 April 2013).

Boyle, F. 1893. *About Orchids: A Chat.* London: Chapman and Hall.

Boyle, F. 1901. *The Woodland Orchids.* London: Macmillan.

Buchmann, S.L. 1987. The ecology of oil flowers and their bees. *Ann. Rev. Ecol.* 18: 343–369.

Burns-Balogh, P., H. Robinson, and M.S. Foster. 1985. The capitate-flowered epiphytic Spiranthinae (Orchidaceae) and a new genus from Paraguay. *Brittonia* 37(2): 154–162.

Calmon, P. 1975. *História de Dom Pedro II*. Rio de Janeiro: José Olímpio.

Calvo, R.N. 1990. Four-year growth and reproduction of *Cyclopogon cranichoides* (Orchidaceae) in south Florida. *Amer. J. Bot.* 77(6): 736–741.

Carnevali, G., and G.A. Romero. 1991. Orchidaceae Dunstervillorum I: A new *Dryadella* from the Venezuelan Guayana. *Novon* 1(2): 73–75.

Chase, M.W., and J.S. Pippen. 1988. Seed morphology in the Oncidiinae and related subtribes (Orchidaceae). *Syst. Bot.* 13(3): 313–323.

Cockerell, T.D.A. 1919. Sir Joseph Hooker. *Science* 49(1275): 525–530.

Cribb, P. 2010. The orchid collection and illustrations of Consul Freidrich C. Lehmann. *Lankesteriana* 10(2,3): 9–30.

Curry, K.J., L.M. McDowell, W.S. Judd, and W.L. Stern. 1991. Osmophores, floral feathers, and systematics of *Stanhopea* (Orchidaceae). *Amer. J. Bot.* 78(5): 610–623.

Dafni, A. 1984. Mimicry and deception in pollination. *Ann. Rev. Ecol. Syst.* 15:259–278.

Dalström, S. 2001. A synopsis of the genus *Cyrtochilum* (Orchidaceae; Oncidiinae): taxonomic reevaluation and new combinations. *Lindleyana* 16(2): 56–80.

Dalström, S. 2003. Orchids smarter than scientists—an approach to Oncidiinae (Orchidaceae) taxonomy. *Lankesteriana* 7:33–36.

Damon, A., and M.A. Pérez-Soriano. 2005. Interaction between ants and orchids in the Soconusco region, Chiapas, Mexico. *Entomotropica* 20(1): 59–65.

Damon, A., and P. Salas-Roblero. 2007. A survey of pollination in remnant orchid populations in Soconusco, Chiapas, Mexico. *Trop. Ecol.* 48(1): 1–14.

Darwin, C. 1862. Letter to J.D. Hooker. Reprinted in Darwin, F. (ed.) 1903. *More Letters of Charles Darwin: A Record of His Work in a Series of Hitherto Unpublished Letters* (Volume 1). London: John Murray.

Darwin, C. 1877. *The Various Contrivances by Which Orchids are Fertilized by Insects*. New York: New York University Press.

Davidse, G. 1989. Obituary of Julian Alfred Steyermark. *Taxon* 38(1): 160–163.

Davies, K.L., and M. Stpiczyńska. 2006. Labellar micromorphology of Bifrenariinae Dressler (Orchidaceae). *Ann. Bot.* 98(6): 1215–1231.

Davies, K.L., and M. Stpiczyńska. 2007. Micromorphology of the labellum and floral spur of *Cryptocentrum* Benth. and *Sepalosaccus* Schltr. (Maxillariinae: Orchidaceae). *Ann. Bot.* 100(4): 797–805.

Davies, K.L., and M. Stpiczyńska. 2008. Labellar micromorphology of two euglossine-pollinated orchid genera; *Scuticaria* Lindl. and *Dichaea* Lindl. *Ann. Bot.* 102:805–824.

Davies, K.L., and M. Stpiczyńska. 2011. Comparative labellar micromorphology of Zygopetalinae (Orchidaceae). *Ann. Bot.* 108(5): 945–964.

Daviña, J.R., M. Grabiele, J.C. Cerutti, D.H. Hojsgaard, R.D. Almada, I.S. Insaurralde, and A.I. Honfi. 2009. Chromosome studies in Orchidaceae from Argentina. *Gen. and Molec. Biol.* 32(4): 811–821.

del Castillo, M., and J.D. Ackerman. 1992. *The Orchids of Puerto Rico and the Virgin Islands*. San José: University of Puerto Rico Press.

de Melo, M.C., E.L. Borba, and E.A.S. Paiva. 2010. Morphological and histological characterization of the osmophores and nectaries of four species of *Acianthera* (Orchidaceae: Pleurothallidinae). *Plant Syst. Evol.* 286:141–151.

Dentinger, B.T.M., and B.A. Roy. 2010. A mushroom by any other name would smell as sweet: *Dracula* orchids. *McIlvainea* 19(1): 1–13.

Dodson, C.H. 1962. Pollination and variation in the subtribe Catasetinae (Orchidaceae). *Ann. Missouri Bot. Gard.* 49(1): 35–56.

Dodson, C.H., and R.J. Gillespie. 1967. *The Biology of Orchids*. Nashville, TN: Mid-America Orchid Congress.

Dodson, C.H. 1993–2004. *Native Ecuadorian Orchids* (Volumes 1–5). Medellín: Editorial Colina.

Dodson, C.H. 2003. Why are there so many orchid species? *Lankesteriana* 7:99–103.

Dodson, C.H., and R.L. Dressler. 1993. *Field Guide to the Orchids of Costa Rica*. Ithaca, NY: Cornell University Press.

Dodson, C.H., and R. Escobar. 1987. The telipogons of Costa Rica (1). *Orquideología* 17(1): 3–69.

Dodson, C.H., and A.H. Gentry. 1991. Biological extinction in western Ecuador. *Ann. Missouri Bot. Gard.* 78(2): 273–295.

Dressler, R.L. 1961. The systematic position of *Cryptocentrum* (Orchidaceae). *Brittonia* 13(3): 266–270.

Dressler, R.L. 1965. Notes on the genus *Govenia* in Mexico (Orchidaceae). *Brittonia* 17(3): 266–277.

Dressler, R.L. 1966. Observations on orchids and euglossine bees in Panama and Costa Rica. *Rev. Biol. Trop.* 15(1): 143–183.

Dressler, R.L. 1967. Pollination by euglossine bees. *Evolution* 22(1): 202–210.

Dressler, R.L. 1968. Notes on *Bletia*. *Brittonia* 20(2): 182–190.

Dressler, R.L. 1971. Dark pollinia in hummingbird-pollinated orchids or do hummingbirds suffer from strabismus? *Amer. Naturalist* 105(941): 80–83.

Dressler, R.L. 1979. *Eulaema bombiformis, E. meriana*, and Müllerian mimicry in related species (Hymenoptera: Apidae). *Biotropica* 11(2): 144–151.

Dressler, R.L. 1981. *The Orchids: Natural History and Classification*. Cambridge, MA: Harvard University Press.

Dressler, R.L. 1982. Biology of the orchid bees (Euglossini). *Ann. Rev. Ecol. Syst.* 13(1): 373–394.

Dressler, R.L. 1993. *Field Guide to the Orchids of Costa Rica and Panama*. Ithaca, NY: Cornell University Press.

Dressler, R.L. 1993. *Phylogeny and Classification of the Orchid Family*. Cambridge, UK: Cambridge University Press.

Dressler, R.L. 2000. Mesoamerican orchid novelties 3. *Novon* 10(3): 193–200.

Dressler, R.L., and W.E. Higgins. 2003. *Guarianthe*, a generic name for the "*Cattleya*" *skinneri* complex. *Lankesteriana* 7:37–38.

Dutra, D., M.E. Kane, C.R. Adams, and L. Richardson. 2009. Reproductive biology of *Cyrtopodium punctatum* in situ: implications for conservation of an endangered Florida orchid. *Plant Species Biol.* 24(2): 92–103.

Ebel, F., and O. Birnbaum. 1972. *The Strange and Beautiful World of Orchids*. Translated by C.S.V. Salt. New York: Van Nostrand.

Endara, L., D.A. Grimaldi, and B.A. Roy. 2010. Lord of the flies: pollination of *Dracula* orchids. *Lankesteriana* 10(1): 1–11.

Fay, M.F., and M.W. Chase. 2009. Orchid biology: from Linnaeus via Darwin to the 21st century. *Ann. Bot.* 104(3): 359–364.

Fenster, C.B., W.S. Armbruster, P. Wilson, M.R. Dudash, and J.D. Thomson. 2004. Pollination syndromes and floral specialization. *Ann. Rev. Ecol. Evol. Syst.* 35:375–403.

Ferry, R.J. 2007. James Bateman and orchid literature. *McAllen International Orchid Society Journal* 8(1): 5–13.

Fisher, B.L. 1992. Facultative ant association benefits a Neotropical orchid. *J. Trop. Ecol.* 8(1): 109–114.

Fisher, B.L., and J.K. Zimmerman. 1988. Ant/orchid associations in the Barro Colorado National Monument, Panama. *Lindleyana* 3(1): 12–16.

Flach, A., R.C. Dondon, R.B. Singer, S. Koehler, M. Amaral, and A.J. Marsaioli. 2004. The chemistry of pollination in selected Brazilian Maxillariinae orchids: floral rewards and fragrance. *J. Chem. Ecol.* 30(5): 1045–1056.

Freiberg, M. 1999. The vascular epiphytes on a *Virola michelii* tree (Myristicaceae) in French Guiana. *Ecotropica* 5:75–81.

Freuler, M.J. 2008. *Orquídeas*. Buenos Aires, Argentina: Editorial Albatros SACI.

Gann, G.D., K.N. Hines, S. Saha, and K.A. Bradley. 2009. *Rare plant monitoring and restoration on Long Pine Key, Everglades National Park: Final Report, Year 5.* Miami, FL: Institute for Regional Conservation.

Garcia-Cruz, J., and V. Sosa. 2005. Phylogenetic relationships and character evolution in *Govenia* (Orchidaceae). *Canadian J. Bot.* 83(10): 1329–1339.

Gravendeel, B., A. Smithson, F.J.W. Slik, and A. Schuiteman. 2004. Epiphytism and pollinator specialization: drivers for orchid diversity? *Phil. Trans. R. Soc. Lond.* 359(1450): 1523–1535.

Hágsater, E. 2005. *Orchids of Mexico.* Mexico City: Productos Farmacéuticos.

Hágsater, E., V. Dumony, and A.M. Pridgeon (IUCN/SSC Orchid Specialist Group). 1996. *Orchids: Status Survey and Conservation Action Plan.* Gland, Switzerland: IUCN.

Hammer, R.L. 1999. The native orchids of Southern Florida. *North American Native Orchid Journal* 5(3): 246–264.

Hansen, E. 1997. The flower of frozen desserts. *Natural History* 106(3): 76–83.

Harding, P.A. 2008. *Huntleyas and Related Orchids.* Portland, OR: Timber Press.

Helbsing, S., M. Riederer, and G. Zotz. 2000. Cuticles of vascular epiphytes: efficient barriers for water loss after stomatal closure? *Ann. Bot.* 86(4): 765–769.

Hietz, P. 1998. Diversity and conservation of epiphytes in a changing environment. *Pure Appl. Chem.* 70(11): 1–11.

Hietz, P., G. Buchberger, and M. Winkler. 2006. Effect of forest disturbance on abundance and distribution of epiphytic bromeliads and orchids. *Ecotropica* 12:103–112.

Hodges, S.A., and M.L. Arnold. 1995. Spurring plant diversification: are floral nectar spurs a key innovation? *Proc. K. Soc. Lond.* 262(1365): 343–348.

Holst, A.W. 1999. *The World of Catasetums.* Portland, OR: Timber Press.

Irvin, R. 1960. The early orchid collectors. *Orchid Review* 68(803).

Jalal, J.S., P. Kumar, and Y.P.S. Pangtey. 2008. Ethnomedicinal orchids of Uttarakhand, Western Himalaya. *Ethnobot. Leaflets* 12(1): 1227–1230.

Jenny, R. 2007. *Cuilauzina pendula*: history and culture of an orchid described in 1825. *Orchids* 76(4): 296–301.

Jersáková, J., S.D., Johnson, and P. Kindlmann. 2006. Mechanisms and evolution of deceptive pollination in orchids. *Biol. Rev.* 81(2): 219–235.

Johannson, D.R. 1977. Epiphytic orchids as parasites of their host trees. *Amer. Orchid Soc. Bull.* 46:703–707.

Johnson, S.D., and T.J. Edwards. 2000. The structure and function of orchid pollinaria. *Plant Syst. Evol.* 222:243–269.

Johnson, T. 1999. *CRC Ethnobotany Desk Reference.* Boca Raton, FL: CRC Press.

Jones, W.E., A.R. Kuehnle, and K. Arumuganathan. 1998. Nuclear DNA content of 26 orchid (Orchidaceae) genera with emphasis on *Dendrobium. Ann. Bot.* 82:189–194.

Kasulo, V., L. Mwabumba, and M. Cry. 2009. A review of edible orchids in Malawi. *J. of Horticulture and Forestry* 1(7): 133–139.

Klumpp, A., and G. Klumpp. 1994. Plants as bioindicators of air pollution at the Serra do Mar near the industrial complex of Cubatão, Brazil. *Env. Pollution* 85(1): 109–116.

Knudsen, J.T., L. Tollsten, and L.G. Bergström. 1993. Floral scents—a checklist of volatile compounds isolated by head-space techniques. *Phytochemistry* 33(2): 253–280.

Koehler, S., and M.C.E. Amaral. 2004. A taxonomic study of the South American genus *Bifrenaria* Lindl. (Orchidaceae). *Brittonia* 56(4): 314–345.

Koopowitz, H. 2001. *Orchids and Their Conservation.* London: B.T. Batsford.

Koopowitz, H. 2005. More on insect vision in flowers. *Orchid Digest* 69(4): 266–268.

Koopowitz, H. 2008. *Tropical Slipper Orchids: Paphiopedilum & Phragmipedium Species & Hybrids.* Portland, OR: Timber Press.

Kowalskowska, A., and H.B. Margońska. 2009. Diversity of labellar micromorphological structures in selected species of Malaxidinae (Orchidales). *Acta Societatis Botanicorum Poloniae* 78(2): 141–150.

Kull, T. 2006. Conservation biology of orchids: introduction to the special issue. *Biol. Cons.* 129(1): 1–3.

Kupper, W., and W. Linsenmaier. 1961. *The Orchids.* Translated by Jean W. Little. Edinburgh, Scotland: Nelson & Sons.

La Croix, I. 2008. *The New Encyclopedia of Orchids: 1500 Species in Cultivation.* Portland, OR: Timber Press.

Laube, S., and G. Zotz. 2006. Neither host-specific nor random: vascular epiphytes on three tree species in a Panamanian lowland forest. *Ann. Bot.* 97(6): 1103–1114.

Lawler, L.J. 1984. Ethnobotany of the Orchidaceae. In J. Arditti (ed.), *Orchid Biology, Reviews and Perspectives*, vol. 3, pp. 27–149. Ithaca, NY: Cornell University Press.

Leitch, I.J., I. Kahandawala, J. Suda, L. Hanson, M.J. Ingrouille, M.W. Chase, and M.F. Fay. 2009. Genome size diversity in orchids: consequences and evolution. *Ann. Bot.* 104(3): 469–481.

Leopardi, C., and L.J. Cumana. 2008. Listado de especies de la familia Orchidaceae para el estado Sucre, Venezuela. *Lankesteriana* 8(1): 93–103.

Lopez, R.G., and E.S. Runkle. 2005. Environmental physiology of growth and flowering of orchids. *HortScience* 40(7): 1969–1973.

Martini, P., C. Schlindwein, and A. Montenegro. 2003. Pollination, flower longevity, and reproductive biology of *Gongora quinquenervis* Ruíz and Pavón (Orchidaceae) in an Atlantic forest fragment of Pernambuco, Brazil. *Plant Biol.* 5(5): 495–503.

Mata-Rosas, M., and V.M. Salazar-Rojas. 2009. Propagation and establishment of three endangered Mexican orchids from protocorms. *HortScience* 44(5): 1395–1399.

Mayda, M. del C., and J.D. Ackerman. 1992. *The Orchids of Puerto Rico and the Virgin Islands.* San Juan, Puerto Rico: University of Puerto Rico.

McCartney, C. 2011. *Encyclia tampensis. Orchids* 80(7): 434–437.

McCormick, M.K., D.F. Whigham, and J. O'Neill. 2004. Mycorrhizal diversity in photosynthetic terrestrial orchids. *New Phytologist* 163(2): 425–438.

McCormick, M.K, D.L. Taylor, K. Juhaszova, R.K. Burnett, D.F. Whigham, and J.P. O'Neill. 2012. Limitations on orchid recruitment: not a simple picture. *Molecular Ecology* 21(6): 1511–1523.

Meinzer, F., and G. Goldstein. 1985. Some consequences of leaf pubescence in the Andean giant rosette plant *Espeletia timotensis. Ecology* 66(2): 512–520.

Meisel, J.E., and C.L. Woodward. 2005. Andean orchid conservation and the role of private lands: a case study from Ecuador. *Selbyana* 26(1,2): 49–57.

Menezes, L.C. 2008. Disappearing habitat of *Cattleya nobilior* and *Cattleya walkeriana* in Brazil. *Orchid Digest* 72(3): 141–143.

Micheneau, C., S.D. Johnson, and M.F. Fay. Orchid pollination from Darwin to the present day. *Bot. J. of the Linnean Society* 161(1): 1–19.

Miller, M.A. 1978. Orchids of Economic Use. *American Orchid Society Bulletin* 28:157–162, 268–271, 351–354.

Millican, A. 1891. *Travels and Adventures of an Orchid Hunter.* London: Cassell & Company.

Mondragón, D. 2009. Population viability analysis for *Guarianthe aurantiaca*, an ornamental epiphytic orchid harvested in Southeast México. *Plant Species Biol.* 24(1): 35–41.

Mondragón, D., A. Santos-Moreno, and A. Damon. 2009. Epiphyte diversity on coffee bushes: a management question? *J. Sustainable Agriculture* 33(7): 703–715.

Mondragón-Palomino, M., and G. Theißen. 2009. Why are orchid flowers so diverse? Reduction of evolutionary constraints by paralogues of class B floral homeotic genes. *Annals of Botany* 104:583–594.

Mori, S.A., and F.C. Ferreira. 1987. A distinguished Brazilian botanist, João Barbosa Rodrigues (1842–1909). *Brittonia* 39(1): 73–85.

Murren, C.J., and A.M. Ellison. 1996. Effects of habitat, plant size, and floral display on male and female reproductive success of the neotropical orchid *Brassavola nodosa*. *Biotropica* 28(1): 30–41.

Neubig, K.M, N.H. Williams, M. Whitten, and F. Pupulin. 2009. Molecular phylogenetics and the evolution of fruit and leaf morphology of *Dichaea* (Orchidaceae: Zygopetalinae). *Ann. Bot.* 104(3): 457–467.

Ng, C.K.Y., and C.S. Hew. 2000. Orchid pseudobulbs—'false' bulb with a genuine importance in orchid growth and survival! *Scientia Horticulturae* 83(3): 165–172.

Nicholson, C.C., J.W. Bales, J.E. Palmer-Fortune, and R.G. Nicholson. 2008. Darwin's beetrap. *Plant Signaling and Behavior* 3(1): 19–23.

Nilsson, L.A. 1988. The evolution of flowers with deep corolla tubes. *Nature* 334(6178): 147–149.

Northern, R.T. 1980. *Miniature Orchids.* New York: Van Nostrand Reinhold.

Northern, R.T. 1988. The kingdom of Lilliput: miniature orchids. *American Orchid Society Bulletin* 57(2): 116–122.

Oakeley, H.F. 1999. *Anguloa:* the species, the hybrids and a checklist of Angulocastes. *Orchid Digest* 63(4): 1–31.

Oakeley, H.F. 2008. *Lycaste, Ida and Anguloa: The Essential Guide.* Self-published.

Orlean, S. 1998. *The Orchid Thief: A True Story of Beauty and Obsession.* New York: Ballantine Books. Reprinted 2011, Random House.

Ossenbach, C. 2005. History of orchids in Central America. Part I: From prehispanic times to the independence of the new republics. *Harvard Papers in Botany* 10(2): 183–226.

Ossenbach, C. 2007. History of orchids in Central America. Part II: The new republics— 1821–1870. *Selbyana* 28(2): 169–209.

Ossenbach, C. 2009. Orchids and orchidology in Central America. 500 Years of history. *Lankesteriana* 9(1–2): 1–268.

Pansarin, E.R., V. Bittrich, and M.C.E. Amaral. 2006. At daybreak—reproductive biology and isolating mechanisms of *Cirrheae dependens* (Orchidaceae). *Plant Biol.* 8(4): 494–502.

Pansarin, L.M., E.R. Pansarin, and M. Sazima. 2008. Facultative autogamy in *Cyrtopodium polyphyllum* (Orchidaceae) through a rain-assisted pollination mechanism. *Australian J. Bot.* 56(4): 363–367.

Pansarin, L.M., E.R. Pansarin, and M. Sazima. 2008. Reproductive biology of *Cyropodium polyphyllum* (Orchidaceae): a Cyrtopodiinae pollinated by deceit. *Plant Biol.* 10(5): 650–659.

Peakall, R., A.J. Beattie, and S.H. James. 1987. Pseudocopulation of an orchid by male ants: a test of two hypotheses accounting for the rarity of ant pollination. *Oecologia* 73(4): 522–524.

Peakall, R., and S.H. James. 1989. Outcrossing in an ant pollinated clonal orchid. *Heredity* 62:161–167.

Pemberton, R.W. 2007. Pollination of *Guarianthe skinneri*, and ornamental food deception orchid in southern Florida, by the naturalized orchid bee *Euglossa viridissima*. *Lankesteriana* 7(3): 461–468.

Pemberton, R.W. 2010. Biotic resource needs of specialist orchid pollinators. *Bot. Rev.* 76(2): 275–292.

Pemberton, R.W., and H. Liu. 2008. Potential of invasive and native solitary specialist bee pollinators to help restore the rare cowhorn orchid (*Cyrtopodium punctatum*) in Florida. *Biol. Conservation* 141(7): 1758–1764.

Pemberton, R.W., and G.S. Wheeler. 2006. Orchid bees don't need orchids: evidence from the naturalization of an orchid bee in Florida. *Ecology* 87(8): 1995–2001.

Peña, A., S. Capella, and C. González. 1995. Characterization and identification of the mucilage extracted from orchid bulbs (*Bletia campanulata*) by high temperature capillary gas chromatography. *J. High Resolution Chromatography* 18(11): 713–717.

Pridgeon, A.M. 1982. Diagnostic anatomical characters in the Pleurothallidinae (Orchidaceae). *Amer. J. Bot.* 69(6): 921–938.

Pridgeon, A.M. 2006. *The Illustrated Encyclopedia of Orchids.* Portland, OR: Timber Press.

Pridgeon, A.M., P.J. Cribb, M.W. Chase, and F.N. Rasmussen (eds.). 1999. *Genera Orchidacearum.* Vol. 1. *General introduction, Apostasioideae, Cypripedioideae.* Oxford, UK: Oxford University Press.

Pridgeon, A M., P.J. Cribb, M.W. Chase, and F.N. Rasmussen (eds.). 2001. *Genera Orchidacearum.* Vol. 2. *Orchidoideae* (Part 1). Oxford, UK: Oxford University Press.

Pridgeon, A M., P.J. Cribb, M.W. Chase, and F.N. Rasmussen (eds.). 2003. *Genera Orchidacearum.* Vol. 3. *Orchidoideae* (Part 2), *Vanilloideae.* Oxford, UK: Oxford University Press.

Pridgeon, A M., P.J. Cribb, M.W. Chase, and F.N. Rasmussen (eds.). 2005. *Genera Orchidacearum.* Vol. 4. *Epidendroideae* (Part 1). Oxford, UK: Oxford University Press.

Pridgeon, A M., P.J. Cribb, M.W. Chase, and F.N. Rasmussen (eds.). 2009. *Genera Orchidacearum,* Vol. 5. *Epidendroideae* (Part 2). Oxford, UK: Oxford University Press.

Pridgeon, A.M., R. Solano, and M.W. Chase. 2001. Phylogenetic relationships in Pleurothallidinae (Orchidaceae): combined evidence from nuclear and plastid DNA sequences. *Amer. J. Bot.* 88(12): 2286–2308.

Pupulin, F. 2009. The natural and taxonomic history of *Chondrorhyncha* (Orchidaceae: Zygopetalinae). The 19th World Orchid Conference, Miami, FL.

Pupulin, F., and D. Bogarín. 2005. The genus *Brassia* in Costa Rica: a survey of four species and a new species. *Orchids* 74(3): 202–207.

Pupulin, F., and H. Medina. 2010. Note on the genus *Dresslerella. Orchid Digest* 74(2): 60–67.

Pupulin, F., and G. Merino. 2010. *Comparettia sotoana* (Orchidaceae: Oncidiinae), a new Ecuadorian species. *Lankesteriana* 9(3): 399–402.

Rach, N. 2008. The Genus *Acineta.* http://www.autrevie.com/Stanhopea/Acineta.html (accessed 2 April 2013).

Ramírez, N. 1989. Biología de polinización en una comunidad arbustiva tropical de la Alta Guayana Venezolana. *Biotropica* 21(4): 319–330.

Ramírez, S.R., B. Gravendeel, R.B. Singer, C.R. Marshall, and N.E. Pierce. 2007. Dating the origin of the Orchidaceae from a fossil orchid with its pollinator. *Nature* 448:1042–1045.

Reed, D.M. 1956. James Bateman and his "Orchidaceae of Mexico & Guatemala." *Indiana University Bookman* 1:27–35.

Reinikka, M.A. 1995. *A History of the Orchid.* Portland, OR: Timber Press.

Rigby, R. 2005. Highlights of Orchid History. http://www.duncan.rigby.net/RR/USAOHP1.pdf (accessed 2 April 2013).

Rivas, R., J. Warner, and M. Bermúdez. 1998. Presencia de micorrizas en orquídeas de un jardín botánico neotropical. *Biol. Trop.* 46(2): 211–216.

Rodríguez-Robles, J.A., E.J. Meléndez, and J.D. Ackerman. 1992. Effects of display size, flowering phenology, and nectar availability on effective visitation frequency in *Comparettia falcata* (Orchidaceae). *Amer. J. Bot.* 79(9): 1009–1017.

Rolfe, R.A. 1887. On bigeneric orchid hybrids. *Bot. J. of the Linnean Society* 24(160): 156–170.

Roubik, D.W. 2000. Deceptive orchids with Meliponini as pollinators. *Plant Syst. Evol.* 222:271–279.

Roubik, D.W., and J.D. Ackerman. 1987. Long-term ecology of euglossine orchid-bees (Apidae: Euglossini) in Panama. *Oecologia* 73(3): 321–333.

Saha, A.K., S. Saha, J. Sadle, J. Jiang, M.S. Ross, R.M. Price, L. Sternberg, and K.S. Wendelberger. 2011. Sea level rise and South Florida coastal forests. *Climatic Change* 107(1): 81–108.

Salazar, G.A., L.I. Cabrera, and C. Figueroa. 2011. Molecular phylogenetics, floral convergence and systematics of *Dichromanthus* and *Stenorrhynchos* (Orchidaceae: Spiranthinae). *Bot. J. of the Linnean Society* 167(1): 1–18.

Salazar, G.A., L.I. Cabrera, S. Madriñán, and M.W. Chase. 2009. Phylogenetic relationships of Cranichidinae and Prescottiinae (Orchidaceae, Cranichideae) inferred from plastid and nuclear DNA sequences. *Ann. Bot.* 104(3): 403–416.

Sandoval-Zapotitla, E., and T. Terrazas. 2001. Leaf anatomy of 16 taxa of the *Trichocentrum* clade. *Lindleyana* 16(2): 81–93.

Sargent, R.D. 2004. Floral symmetry affects speciation rates in angiosperms. *Proc. Biol. Sci.* 271(1539): 603–608.

Sauvêtre, P. 2009. *Maxillaria lehmannii* and its namesake. *Orchid Review* 117:200–205.

Sayers, B., H. duPlooy, and B. Adams. 2007. Working together for orchid conservation—the National Botanic Gardens, Glasnevin and Belize Botanic Gardens. *Lankesteriana* 7(1–2): 153–155.

Schemske, D.W. 1980. Evolution of floral display in the orchid *Brassavola nodosa*. *Evolution* 34(3): 489–493.

Schlessman, M.A. 1986. Interpretation of evidence for gender choice in plants. *Amer. Nat.* 128(3): 416–420.

Schultes, R.E. 1988. Ethnopharmacological conservation: a key to progress in medicine. *Suppl. Acta Amazonica* 18(1–2): 393–406.

Schultes, R.E. 1990. Medicinal orchids of the Indians of the Colombian Amazon. *American Orchid Society Bulletin* 59(2): 159–161.

Schultes, R.E., and R.F. Raffauf. 1990. *The Healing Forest: Medicinal and Toxic Plants of the Northwest Amazonia*. Portland, OR: Dioscorides Press.

Schweinfurth, C. 1958–1961. *Orchids of Peru*. Vol. 1–3. Chicago, IL: Chicago Field Museum of Natural History.

Seijo, E.R. 2009. A new natural hybrid from Cuba. *Orchid Digest* 73(2): 110–113.

Sequeira, L. 2006. Richard Evans Schultes. *Biographical Memoirs*. Washington, DC: National Academy of Sciences.

Sheehan, T., and M.R. Sheehan. 1995. *An Illustrated Survey of Orchid Genera*. Portland, OR: Timber Press.

Siegel, C. 2010. The king, the travelers, and the endless orchids. *Orchid Digest* 74(1): 18–33.

Siegel, C. 2012. Chikanda: a tale of orchids, AIDS, and Zambian bologna. *Orchid Digest* 76(3): 138–144.

Silvera, K. 2002. Adaptive radiation of oil-reward compounds among neotropical orchid species (Oncidiinae). M.Sc. Thesis, University of Florida, Gainesville, FL.

Silvera, K., L.S. Santiago, J.C. Cushman, and K. Winter. 2009. Crassulacean acid metabolism and epiphytism linked to adaptive radiations in the Orchidaceae. *Plant Physiology* 149(4): 1838–1847.

Silvera, K., L.S. Santiago, J.C. Cushman, and K. Winter. 2010. The incidence of crassulacean acid metabolism in Orchidaceae derived from carbon isotope ratios: a checklist of the flora of Panama and Costa Rica. *Bot. J. of the Linnean Society* 163:194–222.

Simon, H. 1975. *The Private Lives of Orchids*. Philadelphia, PA: J.B. Lippincott.

Singer, R.B. 2003. Orchid pollination: recent developments from Brazil. *Lankesteriana* 7:111–114.

Singer, R.B., and A.A. Cocucci. 1999. Pollination mechanism in southern Brazilian orchids which are exclusively or mainly pollinated by halictid bees. *Plant Syst. Evol.* 217(1–2): 101–117.

Singer, R.B., and M. Sazima. 1999. The pollination mechanism in the '*Pelexia* alliance' (Orchidaceae: Spiranthinae). *Bot. J. of the Linnean Society* 131(3): 249–262.

Singer, R.B., B. Gravendeel, H. Cross, and S.R. Ramirez. 2008. The use of orchid pollinia or pollinaria for taxonomic identification. *Selbyana* 29(1): 6–19.

Sinn, M. 2006. *Cattleya mossiae*, national flower of Venezuela. *Orchid Digest* 70(1): 36–39.

Solis-Montero, L., A. Flores-Palacios, and A. Cruz-Angón. 2005. Shade-coffee plantations as refuges for tropical wild orchids in Central Veracruz, Mexico. *Cons. Biol.* 19(3): 908–916.

Sosa, V., and T. Platas. 1997. Extinction and persistence of rare orchids in Veracruz, Mexico. *Cons. Biol.* 12(2): 451–455.

Soto Arenas, M.Á. 2003. *Vanilla* (generic treatment). In A.M. Pridgeon, P.J. Cribb, M.W. Chase, and F.N. Rasmussen (eds.), *Genera Orchidacearum*, vol. 3. *Orchidoideae* (Part 2), *Vanilloideae*, pp. 321–334. Oxford, UK: Oxford University Press.

Stern, W.L., W.S. Judd, and B.S. Carlsward. 2003. Systematic and comparative anatomy of Maxillarieae (Orchidaceae), *sans* Oncidiinae. *Bot. J. of the Linnean Society* 144(3): 251–274.

Stireman, J.O., III, J.E. O'Hara, and D.M. Wood. 2006. Tachinidae: evolution, behavior, and ecology. *Ann. Rev. Entomol.* 51:525–555.

Stpiczyńska, M., and K.L. Davies. 2008. Elaiophore structure and oil secretion in flowers of *Oncidium trulliferum* Lindl. and *Ornithophora radicans* (Rchb.f.) Garay & Pabst (Oncidiinae: Orchidaceae). *Ann. Bot.* 101(3): 375–384.

Stpiczyńska, M., K.L. Davies, and A. Gregg. 2007. Elaiophore diversity in three contrasting members of Oncidiinae (Orchidaceae). *Bot. J. of the Linnean Society* 155(1): 135–148.

Swarts, N.D., and K.W. Dixon. 2009. Terrestrial orchid conservation in the age of extinction. *Ann. Bot.* 104(3): 543–556.

Tan, K.H. 2009. Fruit fly pests as pollinators of wild orchids. *Orchid Digest* 73(3): 180–187.

Tan, K.H., R. Nishida, and Y.C. Toong. 2002. Floral synomone of a wild orchid, *Bulbophyllum cheiri*, lures *Bactrocera* fruit flies for pollination. *J. Chem. Ecol.* 28(6): 1161–1172.

Teixeira, S de P., E.L. Borba, and J. Semir. 2004. Lip anatomy and its implications for the pollination mechanisms of *Bulbophyllum* species (Orchidaceae). *Ann. Bot.* 93(5): 499–505.

Thien, L.B. 1969. Mosquito pollination of *Habenaria obtusata*. *Amer. J. Bot.* 56(2): 232–237.

Thoms, B. 2009. New trends in *Bulbophyllum* breeding. *Orchid Digest* 73(3): 146–116.

Torretta, J.P., N.E. Gomiz, S.S. Aliscioni, and M.E. Bello. 2011. Biología reproductiva de *Gomesa bifolia* (Orchidaceae, Cymbidieae, Oncidiinae). *Darwiniana* 49(1): 16–24.

Tremblay, R.L., E. Meléndez-Ackerman, and D. Kapan. 2006. Do epiphytic orchids behave as metapopulations? Evidence from colonization, extinction rates and asynchronous population dynamics. *Biol. Cons.* 129(1): 70–81.

United States Department of Agriculture (USDA). 2009. *Floriculture Crops 2008 Summary*. Agricultural Statistics Board Sp Cr 6–1 (09). Washington, DC: USDA.

van den Berg, C., W.E. Higgins, R.L. Dressler, W.M. Whitten, M.A.S. Arenas, A. Culham, and M.W. Chase. 2000. A phylogenetic analysis of Laeliinae (Orchidaceae) based on sequence data from internal transcribed spacers (ITS) of nuclear ribosomal DNA. *Lindleyana* 15(2): 96–114.

van der Cingel, N.A. 2001. *An Atlas of Orchid Pollination: America, Africa, Asia and Australia*. Brookfield, NY: A.A. Balkema Publishers.

van der Pijl, L., and C.H. Dodson. 1966. *Orchid Flowers: Their Pollination and Evolution*. Miami, FL: University of Miami Press.

van Doorn, W.G. 1997. Effects of pollination on floral attraction and longevity. *J. of Exp. Bot.* 48(314): 1615–1622.

van Doorn, W.G., and E.J. Woltering. 2004. Senescence and programmed cell death: substance or semantics? *J. Exp. Bot.* 55(406): 2147–2153.

Vaz, A.B.M., R.C. Mota, M.R.W. Bomfim, M.L.A. Vieira, C.L. Zani, C.A. Rosa, and L.H. Rosa. 2009. Antimicrobial activity of endophytic fungi associated with Orchidaceae in Brazil. *Can. J. Microbiol.* 55(12): 1381–1391.

Veitch, H.J. 1889. Orchid culture past and present. *Journal of the Royal Horticultural Society* 11:115–130.

Vinson, S.B., G.W. Frankie, and H.J. Williams. 1996. Chemical ecology of bees of the genus *Centris* (Hymenoptera: Apidae). *Florida Entomologist* 79(2): 109–129.

Wells, H.G. 1894. *The Strange Orchid*. First published in Pearson's Magazine, April 1905. Reprinted in Hammond, J. (ed.). 2001. *The Complete Short Stories of H.G. Wells*. London: Orion Publishing. Online at Project Gutenberg Australia, http://gutenberg.net.au (accessed 14 September 2013).

Wells, T.C.E., and J.H. Willems (eds.). 1991. *Population Ecology of Terrestrial Orchids*. The Hague: SPB Academic Publishing.

Whigham, D.F., J.P. O'Neill, H.N. Rasmussen, B.A. Caldwell, and M.K. McCormick. 2006. Seed longevity in terrestrial orchids—potential for persistent *in situ* seed banks. *Biol. Cons.* 129(1): 24–30.

Whitten, W.M., N.H. Williams, W.S. Armbruster, M.A. Battiste, L.Strekowski, and N. Lindquist. 1986. Carvone oxide: an example of convergent evolution in euglossine pollinated plants. *Syst. Bot.* 11(1): 222–228.

Whitten, W.M., N.H. Williams, R.L. Dressler, G. Gerlach, and F. Pupulin. 2005. Generic relationships of Zygopetalinae (Orchidaceae: Cymbidieae): combined molecular evidence. *Lankesteriana* 5(2): 87–107.

Wiemer, A.P., M. Moré, S. Benitez-Vieyra, A.A. Cocucci, R.A. Raguso, and A.N. Sérsic. 2008. A simple floral fragrance and unusual osmophore structure in *Cyclopogon elatus* (Orchidaceae). *Plant Biol.* 11(4): 506–514.

Wildlife Conservation Society. 2001. Eaten As Food, African Orchids Threatened By Illegal Trade. http://www.sciencedaily.com/releases/2001/08/010801081646.htm (accessed 2 April 2013).

Williams, L.O. 1969. *In memoriam*: Charles Herbert Lankester (1879-1969). *Amer. Orchid Soc. Bull.* 38:860–862.

Williams, N.H. 2009. Molecular systematic in Oncidiinae: chaos or stability. Gainesville, FL: University of Florida Laboratory for Orchid Research & Development.

Wilson, D.M, W. Fenical, M. Hay, N. Lindquist, and R. Bolser. 1999. Habenariol, a freshwater feeding deterrent from the aquatic orchid *Habenaria repens* (Orchidaceae). *Phytochemistry* 50(8): 1333–1336.

Withner, C.L. 2007. The genus *Brassavola* revisited. *Orchid Digest* 71(4): 246–255.

Yam, T.W., J. Arditti, and K.M. Cameron. 2009. "The orchids have been a splendid sport"— an alternative look at Charles Darwin's contribution to orchid biology. *Amer. J. Bot.* 96(12): 2128–2154.

Zelenko, H. 2003. *Orchids: The Pictorial Encyclopedia of* Oncidium. New York: ZAI Publications.

Zelenko, H., and P. Bermúdez. 2009. *Orchids Species of Peru*. New York: ZAI Publications.

Zimmerman, J.K., and I.C. Olmsted. 1992. Host tree utilization by vascular epiphytes in a seasonally inundated forest (Tintal) in Mexico. *Biotropica* 24(3): 402–407.

Zotz, G. 2004. How prevalent is crassulacean acid metabolism among vascular epiphytes? *Oecologia* 138:184–192.

Zotz, G., and G. Schmidt. 2006. Population decline in the epiphytic orchid *Aspasia principissa. Biol. Cons.* 129(1): 82–90.

Zotz, G., and H. Ziegler. 1997. The occurrence of crassulacean acid metabolism among vascular epiphytes from Central Panama. *New Phytol.* 137(2): 223–229.

Zuchowski, W. 2007. *Tropical Plants of Costa Rica: A Guide to Native and Exotic Flora.* Ithaca, NY: Cornell University Press.

Index

Page numbers in **bold** refer to locations in "Orchid Genus Accounts," "Illustrated Glossary," or "Where to See Orchids." Page numbers in *italics* indicate figures and photos. Page numbers for terms that occur with great frequency in genus descriptions (e.g., petal, sepal, pseudobulb) are largely restricted to occurrences in the illustrated glossary and subsequent figures. For orchid species that are subject to taxonomic debate, the page number given is for the broadly accepted species name used in the text; alternative names refer to the broadly accepted name (e.g., *Ornithidium aurea*. See *Maxillaria aurea*).

Abra Patricia-Alto Nieva Conservation Area, 231
Acianthera ramosa. See *Pleurothallis ramosa*
Acineta, **58**
  A. antioquiae, 58
  A. chrysantha, 58
  A. superba, 58
aff. (affinis), **56**
Albius, Edmond, 210
alfeñiques, 129
allopatric speciation, 22, 23, 110
Amboro National Park, 219
Ames, Oakes, 77, 80, 187
Angraecum sesquipedale, 67
Anguloa, 50, 58, **59**
  A. clowesii, 59
  A. virginalis, 59
ant
  colonies and nests, 14, 103, 116, 155, 194
  mutualism with orchids, 84, 101–103, 155, 202
  trap jaw (*Odontomachus*), 116
anther, *36*
apex (apical), *41*, *42*
apomixis, 215

aposematic coloration, 60
APROVACA Orchid Conservation Center, 229
Argentina, **218**, 230
Armando Bermudez National Park, 223
Aspasia, 48–49, **60**
  A. epidendroides, 60
  A. principissa, 60
  A. silvana, 60
Atlantic Forest, 101, 134, 219–220
Augusta, Isabel, 124
Augusto Ruschi Biological Reserve, 219
Aztec people, 65, 210

back bulb, 13–14, 71
Bahamas, The, **219**
Barbosella, 51, **61**
  B. cogniauxiana, 61
  B. dolichorhiza, 61
  B. prorepens, 61
Barkeria, 48, 53, **62**
  B. halbingeri, 62
  B. lindleyana, 62
Barro Colorado Island Research Station, 229
Baru Volcano National Park, 229

Lightning Source UK Ltd.
Milton Keynes UK
UKHW021453271120
374153UK00001B/4